ENCYCLOPEDIA
OF
Earth and Physical Sciences

5

Hydrogen – Marine exploration

Marshall Cavendish
New York • London • Toronto • Sydney

Cover illustration: abstract model of an atom
Title page illustration: liquid droplet

Marshall Cavendish Corporation
99 White Plains Road
Tarrytown, New York 10591-9001

© Marshall Cavendish Corporation, 1998

Created by **Brown Packaging Partworks Ltd**

Library of Congress Cataloging-in-Publication Data

Encyclopedia of earth and physical sciences.
        p.   cm.
        Includes bibliographical references and index.
        Contents: v. 1. Abs–Cal—v. 2. Cam–Cry—v. 3. Cur–Ext—v. 4 Fau–Hyd—v. 5. Hyd–Mar—v. 6. Mar–Nov—v. 7. Nuc–Pla—v. 8. Pla–Sil—v. 9. Sil–Tim—v. 10 Tim–Zin—v. 11. Index.
        ISBN 0-7614-0551-8 (set)
        1. Earth sciences—Encyclopedias, Juvenile. 2. Physical sciences—Encyclopedias, Juvenile. [1. Earth sciences—Encyclopedias. 2. Physical sciences—Encyclopedias.] I. Marshall Cavendish Corporation.
QE5.E513   1998
500.2'03—dc21   96-49660
                              CIP
                              AC

Printed in Malaysia
Bound in U.S.A.

PHOTOGRAPHIC CREDITS

Ray Dunn: 626.
Mary Evans Picture Library: 591, 592, 668, 699, 705, 706, 707.
Frank Lane Picture Agency: 586, 590, 598, 601, 608, 623, 641, 648, 650, 652, 675, 676, 694, 718.
Science Photo Library: 585, 587, 588, 593, 594, 599, 600, 602, 603, 604, 606, 607, 609, 610, 612, 613, 615, 616, 617, 620, 621, 622, 624, 625, 628, 630, 633, 634, 636–640, 642, 643, 645, 646, 647, 649, 651, 653, 654, 655, 657–666, 669, 671, 673, 674, 677–681, 683, 685–689, 691, 692, 695, 698, 700–704, 708, 712–716.
T. R. H. Pictures: 581, 582, 605, 684, 717.

ARTWORK CREDITS

Bill Botten: 584, 589, 618, 619, 627, 632, 635, 644, 656, 667, 672, 690.
Jennie Dooge: 582, 593, 596, 597, 611, 613, 614, 629, 631, 635, 660, 696, 697, 678, 682, 698, 693, 670.

CONTENTS

HYDROGEN

With an atomic mass of just 1.00794 and comprising about 90 percent of the atoms in the Universe, hydrogen is the lightest and the most abundant element. Hydrogen atoms have the simplest structure of all the elements, and many scientists consider hydrogen to be the building block from which all other elements are made. Under normal conditions, hydrogen exists as diatomic molecules (H_2). It is a colorless, tasteless, odorless gas at room temperature.

Isotopes of hydrogen

Hydrogen exists in three different forms, or isotopes. These are atoms of the same element with different mass numbers (see ISOTOPES). Hydrogen-1, also known as protium, has the simplest atomic structure of all the elements, with one proton in its nucleus. This is the most common isotope, making up 99.9844 percent of naturally occurring hydrogen.

Hydrogen-2, or deuterium, has one proton and one neutron in its nucleus. Deuterium makes up 0.0156 percent of all the hydrogen in nature.

Tritium is the name given to hydrogen-3, whose nucleus holds one proton and two neutrons. It constitutes 10^{-15} to 10^{-16} percent of all naturally occurring hydrogen. Tritium is radioactive with a half-life of 12.3 years. It is formed in Earth's upper atmosphere (see ATMOSPHERE) and only exists in trace amounts.

The physical properties of the three hydrogen isotopes vary (see the table below). The chemical properties of deuterium, however, are similar to those of protium, though deuterium is slightly less reactive. This is due, in part, to the fact that at any given temperature, deuterium atoms collide less frequently than do protium atoms, so chemical reactions proceed more slowly (see CHEMICAL REACTIONS). For the same reason, tritium is even less reactive than deuterium.

Ever since the existence of isotopes had been postulated by English chemist Frederick Soddy (1877–1956) in 1913, scientists had speculated on the existence of isotopes of hydrogen. Then, in 1931, U.S. chemist Harold Urey (1893–1981) reasoned that if heavier isotopes of hydrogen existed, they would

Hydrogen was responsible for the explosion of the **Hindenburg** *airship in 1937. The gas is less dense than air, but it is also highly flammable.*

have a lower vapor pressure than ordinary hydrogen. This meant that if a mixture of hydrogen isotopes was liquefied and then vaporized, ordinary hydrogen would vaporize before the heavier isotopes.

Based on this reasoning, Urey performed a number of experiments with liquefied hydrogen and analyzed the fractions of the liquid that vaporized and stayed behind. The analysis revealed the presence of deuterium in the fraction that remained. Urey had discovered deuterium, and this achievement earned him the Nobel Prize for chemistry in 1934.

The discovery of tritium—really its invention—was made by Australian physicist Marcus Oliphant. In 1934, Oliphant bombarded atoms of deuterium with deuterium nuclei, or deuterons. The product was a still heavier isotope of hydrogen, tritium. It was not until 1946 that U.S. chemist Willard Libby (1908–1980) showed that tritium was produced naturally in the upper atmosphere by the impact of

CONNECTIONS

● Hydrogen has three **ISOTOPES**.

● Hydrogen bonding is responsible for many of the properties that **WATER** possesses as a **LIQUID**.

CORE FACTS

■ Hydrogen is the lightest and the most abundant element in the Universe.

■ Hydrogen exists in three forms or isotopes: protium, deuterium, and tritium.

■ Although hydrogen gas (H_2) is relatively inert, atomic hydrogen (H) is highly reactive and is a powerful reducing agent.

■ Hydrogen is prepared industrially by reacting calcium hydride with water or by treating hydrocarbons with steam.

SELECTED PHYSICAL PROPERTIES OF HYDROGEN ISOTOPES AND THEIR WATERS

Property	Protium	Deuterium	Tritium
Melting point (°F)	-434.56	-425.97	-421.57
Melting point (°C)	-259.2	-254.43	-251.98
Boiling point (°F)	-422.99	-417.08	-414.62
Boiling point (°C)	-252.77	-249.49	-248.12
Heat of vaporization (cal/mole)	216	293	333

THE HYDROGEN FUEL CELL

The Apollo space program that took humans to the Moon and back was made possible in part by hydrogen fuel cells. These cells produced electricity for use by the Apollo space vehicles, much as conventional batteries might. The key difference was that the electricity-producing components of Apollo's "batteries" were not zinc and copper sulfate, but hydrogen and oxygen. In addition, the fuel cells weighed about 500 lb (225 kg), rather than the few tons that a conventional battery of similar electrical capacity might weigh.

In a hydrogen fuel cell, chemical energy is converted directly into electrical energy. This occurs when hydrogen fuel is oxidized by oxygen piped into the cell. Hydrogen is fed to one electrode in the cell, while oxygen is fed to the other electrode. The electrodes are separated by an electrolyte, such as an aqueous solution of potassium hydroxide.

At the anode, hydrogen gas combines with hydroxide ions to form water and free electrons. The electrons flow to the cathode, where they combine with oxygen gas and water to reconstitute the hydroxide ions. So long as hydrogen and oxygen are supplied, the fuel cells will continue to produce electricity.

Apollo fuel cells were designed to operate for 11 days. In the case of the ill-fated *Apollo-13* mission, the accidental loss of oxygen reserves drastically reduced the life of the mission's fuel cells. If it were not for the heroic measures taken on the ground and in space, the mission might have failed completely and the astronauts might have been lost (see APOLLO MISSIONS). Fortunately, the measures taken to conserve electricity and oxygen allowed the astronauts to return safely to Earth.

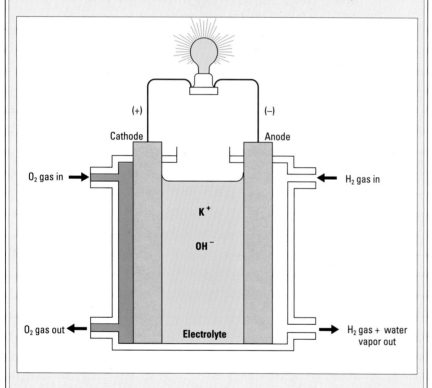

In a hydrogen-oxygen fuel cell, chemical energy is converted to electrical energy through the reaction of hydrogen with oxygen.

A CLOSER LOOK

cosmic radiation on nitrogen to form neutrons, which then react with more nitrogen atoms to generate tritium (see COSMIC RADIATION).

Chemical properties

At room temperature, hydrogen gas (H_2) is relatively unreactive, but its reactivity increases as the temperature rises. Atomic hydrogen (H), however, is very reactive and is a powerful reducing agent. It reacts directly with various metals to produce hydrides and with water to produce hydrogen peroxide (H_2O_2). It is also used to reduce the oxide of some metals in a process that liberates the metals themselves.

Hydrogen combines with various organic compounds to produce others. For example, the hydrocarbons ethane and butane are products of the reaction of hydrogen with ethylene. Important halogen acids, such as hydrochloric acid (HCl), are produced by reacting hydrogen with halogens (see HALOGENS). Ammonia (NH_3) is industrially manufactured by the reaction of a mixture of carbon monoxide and hydrogen with nitrogen.

Hydrogen also reacts with oxygen to produce water, and the cohesive properties of the resulting molecules (that is, their tendency to clump together) are due to hydrogen bonding. This is the affinity of the positive hydrogen "ends" of one water molecule for the negative oxygen part of other nearby water molecules (see CHEMICAL BONDS). Hydrogen bonding also contributes to the properties of other hydrogen-containing compounds, such as ammonia and hydrogen fluoride.

At room temperature, the reaction between hydrogen and oxygen is very slow, but if the reactants are heated, the reaction becomes so fast that it produces an explosion. Such an explosion destroyed the hydrogen-inflated airship *Hindenburg* over Lakehurst, New Jersey, in 1937. Under controlled conditions, the reaction of hydrogen and oxygen can be beneficial—when hydrogen is used as rocket fuel, for example (see the box at right).

Preparation of hydrogen

Hydrogen can be made in the laboratory by reacting an acid with certain metals. In such a reaction, the metal displaces the hydrogen in the acid and the hydrogen gas bubbles off. A typical reaction of this kind involves reacting zinc with hydrochloric acid. In this case, the products are zinc chloride and hydrogen gas:

$$Zn + 2HCl \rightarrow ZnCl_2 + H_2$$

Large quantities of hydrogen are prepared industrially using a number of reactions. These include the reaction of calcium hydride with water and the treatment of volatile hydrocarbons with steam in the presence of a nickel catalyst.

Acids

One of the earliest definitions of an acid is a substance that releases hydrogen ions when it is dissolved in water. This definition was based on a theory provided by Swedish chemist Svante Arrhenius (1859–1927) in 1884. It helped explain the properties of acids, especially their ability to conduct electricity.

Arrhenius proposed that electrically charged particles existed in certain solutions. In the case of acids, one of these particles turned out to be the hydrogen ion (H^+), and Arrhenius defined an acid as a water

HYDROGEN AND THE BIRTH OF THE UNIVERSE

Scientists believe that during the first few minutes of the Universe's existence—when temperatures were between 100 million and 3 billion degrees Kelvin—deuterium nuclei fused to produce an isotope of helium (helium-3; see HELIUM) and tritium. Two helium-3 nuclei then fused to make helium-4. Next, a helium-4 nucleus and a helium-3 nucleus fused to produce beryllium-7. Then the nucleus of beryllium-7 captured an electron to become lithium-7. Later, within the core of growing stars, other nuclear reactions occurred to form the other elements or their precursors.

Thus, all of the elements in the Universe owe their existence to a single building block—the nucleus of a hydrogen atom.

The Universe contains many different elements, but the origin of all of them is the nucleus of one element: hydrogen.

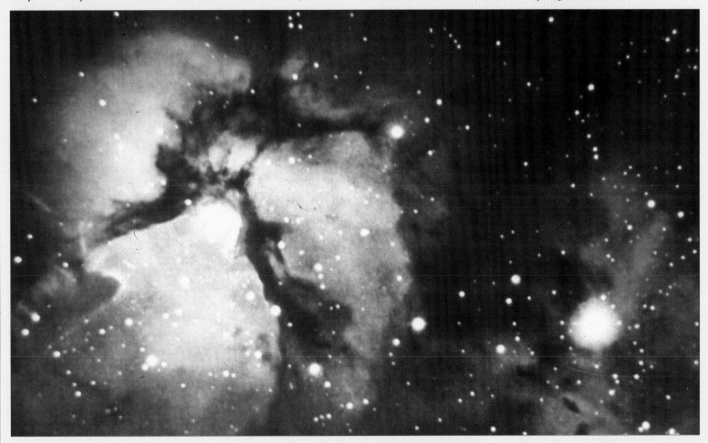

HISTORY OF SCIENCE

solution that produced an excess of these ions. Today, we know that these hydrogen ions can react with water to form a hydronium ion (H_3O^+).

Regardless of whether the hydrogen or hydronium ion exists in water, one thing is certain: an acid is a substance that when dissolved donates a proton (hydrogen ion) to the solution. This proton is responsible for the chemical properties of the acid, which allow acids to be involved in the fundamental reaction called the oxidation-reduction or redox reaction.

C. PROUJAN

See also: ACIDS AND BASES; ELEMENTS; HYDROCARBONS; ISOTOPES; OXIDATION-REDUCTION REACTIONS.

Further reading:
Barrett, J. R. *Understanding Inorganic Chemistry*. New York: Ellis Horwood, 1991.
Cotton, F. A. *Basic Inorganic Chemistry*. New York: John Wiley & Sons, 1995.

HEAVY WATER

Heavy water is water in which the oxygen atoms combine not only with protium, but with deuterium (D_2). It is mainly used in nuclear reactors, where it is a moderator. This means that the heavy water helps control the rate of nuclear reactions in the reactor.

Heavy water is also used in biological research. For example, certain simple organisms such as algae, bacteria, and molds are grown in cultures where the only source of water is heavy water. As a result, all the hydrogen-containing substances produced by these organisms contain deuterium. Such substances are used in special research projects and analytical procedures where the deuterium is used as a label to trace or analyze the reactants or products of biological processes. Heavy water cannot be used in a similar way in animals and higher plants. In fact, in the body fluids of animals, a concentration of deuterium in excess of one-third the total concentration of hydrogen is deadly. In higher plants, two-thirds of deuterium is deadly.

SCIENCE AND SOCIETY

HYDROLOGY

Water is continually circulating between the atmosphere, the ground surface, and the region below ground.

CONNECTIONS

● Atmospheric **WATER** is returned to the surface as **RAIN, SLEET, AND SNOW.**

● Solid **ROCKS** that contain a number of small holes, or pores, may allow the through-flow of **GROUNDWATER.**

On Earth, water is present in all its phases. There is liquid water in streams, lakes, seas, and below the ground. There is solid ice packed into glaciers and ice fields. There is also water vapor in the atmosphere.

The science of hydrology includes the study of water in its various forms: groundwater, surface water, and atmospheric water. It also includes the mechanisms by which water moves over and under Earth, how it changes between forms (the processes of evaporation, precipitation, infiltration, groundwater flow, runoff, and stream flow), and the transport of materials dissolved or suspended in the water.

The hydrologic cycle

Water on Earth circulates through a variety of environments, from high in the atmosphere to deep below the surface. The water-containing region around Earth is called the hydrosphere. The movement of water through the hydrosphere is called the hydrologic cycle. It is an unending solar-powered process. The heat from the Sun transforms liquid water into gaseous water vapor by evaporation. In the atmosphere, water vapor is carried by the winds to every part of the globe. Sometimes this vapor is visible as clouds, although 94 percent of atmospheric water is invisible to the human eye.

When atmospheric water can no longer be carried by the atmosphere, it is returned to the surface

CORE FACTS

■ The hydrologic cycle includes Earth's liquid water, solid ice, and water vapor.

■ The hydrosphere, the water-bearing region that surrounds Earth, extends 9⅓ miles (15 km) above Earth's surface and at least 3 miles (4.8 km) below.

■ The oceans and seas hold more than 97 percent of Earth's water, and more than 2 percent is locked up in glaciers and ice fields.

■ Only 0.625 percent of Earth's water is available as fresh water.

■ Fresh water is essential for the survival of humans and the world's land animals and plants, so the study of hydrology is vitally important.

as precipitation; that is, as dew, rain, sleet, or snow. On the surface, it can be ingested by animals or taken up by plants and then transpired back into the atmosphere. It can infiltrate below the surface of the Earth to recharge groundwater. It can also remain on the surface as runoff in streams or lie frozen as glacial ice. Eventually, in the waters of lakes, streams, or oceans, the water evaporates once more to complete the hydrologic cycle.

Water distribution

Water dominates the face of Earth. The oceans cover 70 percent of Earth's surface and hold about 97.212 percent of the water. Around 2.15 percent is locked up in glaciers and ice caps and about 0.62 percent is hidden below Earth's surface as groundwater. Inland seas and saline lakes contain 0.008 percent and surface water (streams, marshes, and lakes) accounts for 0.009 percent. A tiny 0.001 percent of all Earth's water is in the atmosphere. But, because the atmosphere is where the effects of the Sun's heat are most dramatically felt, the contribution of atmospheric water to the action of the hydrological cycle is greatly out of proportion to its quantity.

Atmospheric water is also disproportionately important to life on Earth because of the speed with which it cycles to another form. From the time a water molecule is evaporated and becomes a part of the atmosphere until the time it condenses out as precipitation, an average of 10 days elapses. The average time—the "residence time"—for water in river channels is about two weeks; for soil moisture, two weeks to one year; for swamps, lakes, and reservoirs, up to 10 years; for glaciers and ice caps, 10 to 1000 years; for oceans and seas, 4000 years; and for groundwater, two weeks to 10,000 years.

Residence time means a great deal to humans, who depend on the tiny 3 percent of water on Earth that is fresh. Because about 68 percent of all fresh water is locked up in glaciers for hundreds of years, only a small portion of it is readily available surface water. Because it may take up to 10,000 years to recharge an aquifer (a water-bearing geological formation) after it has been exhausted, groundwater is especially vulnerable to overuse and pollution. However, groundwater is becoming an ever more important source of fresh water as surface supplies grow more limited.

Groundwater

Water that sinks below Earth's surface is called groundwater. It is found under much of Earth, sometimes very deep down. Most groundwater does not flow in clearly defined streams, except in relatively rare cases, such as streams flowing through limestone caverns. Groundwater flows through Earth's pores, even through apparently solid rock. The number of pores in a material (its porosity) and their arrangement (its permeability) determine how much water can be held by and how fast it can flow through the material. The upper limit of groundwater, called the water table, can vary from place to

place, depending on the amount of water that infiltrates from the surface, the amount of porous material available to hold it, and the amount discharged in springs, oases, or swamps.

An aquifer is the area of rock and earth saturated with groundwater. It is porous material situated on top of an impermeable layer, such as clay or bedrock. The water table closely follows the contour of the land, and water within the aquifer will generally flow under the force of gravity toward lower levels, although sometimes pressure will force the water up to a higher level. When the natural flow is through a material that can be dissolved by the slightly acidic groundwater, such as limestone, erosion will carve out caverns.

Surface water

Surface water is the most visible portion of the hydrologic cycle. It includes the oceans, glaciers and ice sheets, reservoirs, inland seas and lakes, and rivers and streams.

In this pond, wastewater from paper mills is being cleaned. Toxic pollutants are removed by sedimentation, filtration, and bacterial fermentation.

Drought, such as this one affecting the area around a river in the Kruger National Park, South Africa, is a natural extreme of the water cycle.

The oceans contain approximately 320 million cubic miles (1.34 billion km^3) of water, by far the greatest amount of all the water bodies. Glaciers and ice sheets hold about 7 million cubic miles (29 million km^3). Freshwater lakes hold 22,000 cubic miles (91,000 km^3), while saltwater lakes hold 20,500 cubic miles (85,400 km^3). The world's marshes contain around 2750 cubic miles (11,470 km^3) and flowing rivers and streams hold only about 509 cubic miles (2120 km^3).

Surface water is the main agent for weathering rocks and soil, breaking them down into sediments and dissolved ions, and carrying them to the sea. It is also the water on which most living species depend.

Humans are included among these dependent species. For household and industrial consumption, most cities draw surface water from the nearest river or lake or from human-made storage reservoirs. They divert water for irrigation of farmland, for the production of electricity, and for transporting goods.

HUMANS AND THE HYDROLOGIC CYCLE

Humans cannot avoid interfering in the hydrologic cycle, and human use of water often changes it. When people have completed irrigation, consumption, or wastewater treatment, the water is returned to a different place in the hydrologic cycle and is generally lower in quality.

Because groundwater makes up such a large percentage of fresh water that is not locked up in ice, particularly in arid regions, humans pump it for their own use. Pumping more groundwater than can be naturally recharged lowers the water table. This damages dependent plants, allowing saltwater intrusion in coastal areas. Dams and other diversions interfere with the natural flow of sediments into the sea. Draining of wetlands for farms and other development changes the wildlife habitat. It also affects the natural flood control and water purification properties of these areas.

Worldwide, one of the greatest impacts humans have had on the hydrologic cycle has been through tropical deforestation. During such deforestation, the cutting of trees brings not only increased erosion and silting of water bodies, but also causes changes to local climates and even global climate. More than half the rainfall in tropical forests is forest-generated; being caused by evaporation from the forests and altered air patterns over the forests. Extensive deforestation will therefore lead to changes in Earth's atmosphere and, consequently, to precipitation.

It is the role of the hydrologist to study and understand these changes so that appropriate policy decisions governing water use can be made.

A CLOSER LOOK

They also try to control the natural extremes of the hydrological cycle, such as floods and droughts.

Through using water, humans alter the hydrologic cycle. They change how much water evaporates and how much is allowed to recharge the groundwater they have used. They alter the chemical composition of the water by introducing pesticides and wastes that affect the oxygen-carrying capability of water and thus its ability to support fish and other wildlife. All these effects are part of the study and application of hydrology.

Precipitation

The process by which water in the atmosphere is returned to the surface is called precipitation. The behavior of water in the atmosphere, although part of the hydrologic cycle, is generally studied in the separate science of meteorology (see METEOROLOGY).

Precipitation occurs when moist air is cooled. Precipitation can take different forms, and these are determined by temperature, wind, instability of the atmosphere, and the amount of moisture the air contains (see RAIN, SLEET, AND SNOW). There are various types of precipitation:
• **Dew:** Droplets of water that condense from air adjacent to a surface when the air temperature falls below the dew point.
• **Frost:** Frozen dew that forms when the surrounding air temperature falls below the freezing point of water, 32°F (0°C).
• **Fog:** A layer of visible water vapor—a cloud—near the ground that is formed from millions of tiny floating water droplets.
• **Rime:** A granular accumulation of frozen fog.
• **Drizzle:** Liquid precipitation in droplets less than 1/50 in (0.5 mm) in diameter.
• **Rain:** Liquid precipitation in drops larger than 1/50 in (0.5 mm) in diameter.
• **Sleet:** Rain that freezes into ice pellets by falling through a layer of air that is below freezing point.
• **Hail:** Lumps of ice created in cumulonimbus clouds (thunderheads) when rain droplets freeze (see CLOUDS). The droplets are carried up into the cloud by strong updrafts, making even more ice accumulate. Hailstones fall when they become too heavy to be lifted back into the cloud.
• **Snow:** Precipitation that has frozen in the atmosphere as hexagonal ice crystals. These may fall individually or in clusters called snowflakes. Snow is a critical element in the supply of surface water. In many regions of North America, for example, particularly in the western states, snowmelt supplies most of the water in the summer. In California, as much as 90 percent of annual stream flow is from snow that fell during the few months of winter.

Most of the world's fresh water originally fell as snow and remains frozen for hundreds of years in the snowfields, glaciers, ice sheets, and ice caps of the colder regions of the planet. It does not take much of a fall in global temperatures to keep this water frozen. During the last ice age, which ended around 10,000 years ago, 33 percent of the planet's land surface was covered with ice, compared to about 12 percent today.

J. RHODES

See also: GLACIERS; ICE; LAKES; OCEANS AND OCEAN ABYSSES; RAIN, SLEET, AND SNOW; RIVERS; WATER.

Further reading:
Dingman, S. L. *Physical Hydrology*. New York: Macmillan, 1994.
Handbook of Hydrology. Edited by D. Maidment. New York: McGraw-Hill, 1993.
Shaw, E. M. *Hydrology in Practice*. New York: Chapman & Hall, 1994.

POLLUTION OF WATER BODIES

In some parts of Canada, downwind from coal-fired electrical power plants in the United States, the rain that falls is as acidic as the juice from a lemon. Acid rain is formed when smoke containing sulfur dioxide from coal combustion combines with atmospheric oxygen. Dilute sulfuric acid falls to the ground with the rain, and this endangers plant and animal life.

But acid rain is only one of the harmful effects humans have had on surface water. In every corner of human habitation, from mud village huts to the skyscrapers of Manhattan, water is degraded by use and abuse.

In the cities, pollution comes from industrial and household chemicals, landfills, mines, oil fields and underground petroleum storage tanks, and sewage sludge and septic systems. In rural areas, farmers use more and more pesticides and fertilizers, which leach into groundwater and streams. The use of helicopters to spread these chemicals (see below) means that large areas can be affected.

Many areas of the world, particularly those with the most rapidly growing populations, place enormous demands on the water supply. It is an even greater challenge to keep it suitable for consumption. This is a grave public health issue, since many of the most feared diseases, such as cholera, are linked to poor hygiene that is the result of a lack of sufficient supplies of clean water.

SCIENCE AND SOCIETY

HYDROTHERMAL VENTS

Hydrothermal vents are volcanic formations on the seafloor that release heated, mineral-rich water

Communities of fish, crustaceans, and bivalves congregate around a hydrothermal vent in the floor of the Pacific Ocean.

CONNECTIONS

● **VOLCANOES** are vents from which igneous **MATTER**, solid **ROCK** debris, and **GASES** are erupted.

● Organisms that live near vents are similar to those found early in the **FOSSIL RECORD**.

At hydrothermal vents, the heat of Earth's interior meets the frigid water of the deep sea. Like volcanoes on land, which can expel clouds of ash and gas into the atmosphere, hydrothermal vents spew hot, mineral-rich water into the ocean. Hydrothermal vent discharges are supplied by ocean water that has seeped through Earth's crust. The water seeps through fractures in Earth's crust until it reaches molten rock, or magma. The heat of the molten rock causes the water to rise rapidly to the surface, where it is expelled at hydrothermal vents. As the vent fluid cools, minerals in the water precipitate to form dramatic structures—such as chimneys, mounds, or ledges—around the vents.

Hydrothermal vents are found thousands of feet below the ocean surface along deep-sea spreading ridges, which mark the boundaries between tectonic plates (see PLATE TECTONICS).

Vents have been discovered in a number of locations, including the Galápagos Rift in the Pacific Ocean near the equator, the East Pacific Rise and the Juan de Fuca Ridge in the North Pacific, and the Mid-Atlantic Ridge in the North Atlantic.

Physical properties

Temperatures recorded in vent fluids have ranged from about 45°F (7°C) to as high as 750°F (400°C).

At depth, typical temperatures for ocean water are approximately 35°F (1.6°C). The fluid expelled through vents stays liquid at temperatures above the boiling point 212°F (100°C), because intense pressure at depth keeps the water from boiling. Most vent fluids are slightly acidic and are usually oxygen depleted. This means that the fluid is in a chemically reduced state, which is important because it means that certain minerals (for example, metal sulfides) stay in dissolved form until the reduced fluid mixes with oxygen-rich seawater and oxidation reactions cause solid mineral crystals to precipitate.

The chemical composition of seawater also changes as it travels through Earth's crust. Chemicals, such as sulfates and magnesium, precipitate out of the water and are left in the crust as the result of high-temperature and high-pressure chemical reactions

CORE FACTS

■ The chemical composition of seawater changes as it travels through Earth's crust.
■ Due to the high mineral content, vent fluids are slightly acidic and are usually oxygen depleted.
■ Species unique to these hot, sulfur-rich, oxygen-depleted environments exist as part of a complex food chain.

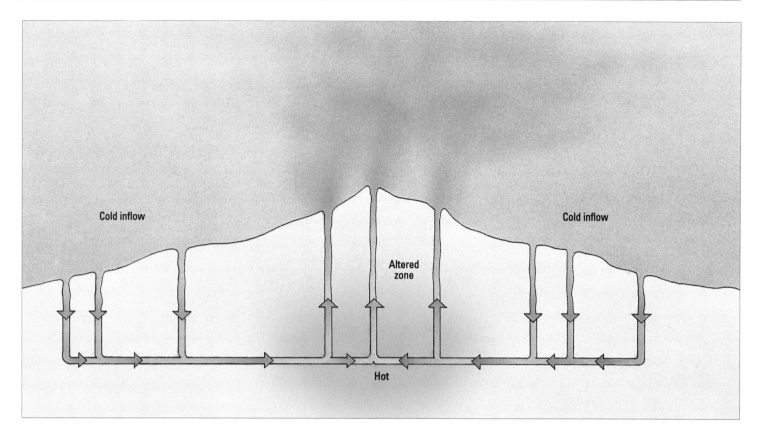

Cold inflow

Cold inflow

Altered zone

Hot

between basalt and seawater; in turn, the seawater leaches other elements from the hot rock. Hydrogen sulfide, a toxic compound that smells like rotten eggs, is one of the most common compounds found in vent plumes. Other compounds include iron, zinc, manganese, and copper sulfides. The vents also contain trace amounts of silver and gold.

Some of the most dramatic formations are chimneys called black smokers, named for the dark plumes of fluid that spew from them. The color of the plumes and of the chimneys themselves comes primarily from fine particles of iron sulfide. Black smokers often measure more than 33 ft (10 m) high and 13 ft (4 m) wide. White smokers also exist, from which issue fluids rich in barium-containing minerals. Vent-related mineral deposits can also take the shape of mounds or shelflike projections called flanges.

Food chains

Bacteria-like organisms called archaebacteria form the base of the food web at hydrothermal vents. Archaebacteria manufacture food by chemosynthesis, using the energy liberated in chemical reactions involving vent fluids, particularly the oxidation of sulfide compounds.

Some organisms feed directly on the archaebacteria. Others have formed symbiotic (mutually beneficial) relationships with them. For example, in its adult form, the thick-plumed rift worm *Riftia pachyptila*, found at the Galápagos Rift, does not possess a digestive system; the tube worm's body is a host to the archaebacteria, which supply the worm with food. Other organisms found at the Galápagos Rift include giant clams (*Calyptogena* spp.), mussels (*Bathymodiolus* spp.), and crabs (*Bythograea* spp.).

K. FREEMAN

See also: OCEANS AND OCEAN ABYSSES; PLATE TECTONICS; VOLCANOES.

Further reading:

Cone, J. *Fire Under the Sea: The Discovery of the Most Extraordinary Environment on Earth—Volcanic Hot Springs on the Ocean Floor.* New York: Morrow, 1991.
Lalli, C. M. *Biological Oceanography: An Introduction.* New York: Pergamon Press, 1993.
Hydrothermal Vents and Processes. Edited by L. M. Parson, C. L. Walker, and D. R. Dixon. London: Geological Society, 1995.

Ocean water seeps into fractures deep in Earth's crust until it reaches hot magma. The molten rock heats the water and alters it chemically. The hot water quickly rises to the surface, emerging at the hydrothermal vents.

THE DISCOVERY OF HYDROTHERMAL VENTS

In February 1977, an oceanographic team led by Robert Ballard of the Woods Hole Oceanographic Institution and Jack Corliss of Oregon State University began searching for deep-sea volcanic formations on the Galápagos Rift off Ecuador—an area where temperature anomalies at great depths, known as temperature spikes, had been detected in the water. Ordinarily temperatures decrease with depth in the ocean, and scientists were surprised to find unusually warm water at certain depths. The Ballard/Corliss team deployed an underwater camera sled called ANGUS. ANGUS shot thousands of frames of slate-blue seafloor and a few images of clams and mussels in silty water. Corliss suspected that the temperature spikes and the location of the shellfish were related. On February 19, he descended 9200 ft (2800 m) to the seafloor in the tiny submersible *Alvin.* After cruising through a nearly lifeless seascape of cliffs, crevasses, and lava flows, the scientists found an oasis crowded with clusters of giant clams and tube worms projecting from cracks in the rocks. The water around the organisms seemed to shimmer and was 10 to 15 degrees warmer than the near-freezing ocean water around it. During the four-week expedition, the team discovered and mapped five hot springs in the Galápagos Rift.

DISCOVERERS

ICE

The hydrogen bonding and small distance between molecules in the crystal structure of a mass of ice leaves little room for inorganic ions or organic molecules. For this reason, sea ice is almost free of salt, except where it is contaminated by contact with seawater.

CONNECTIONS

● **GLACIERS** are ice masses that result from the compacted snow that accumulates in hollows.

● **PERMAFROST** forms where the air temperature is sufficiently cold to keep **GROUNDWATER** permanently frozen.

There is only one common substance that is familiar to us in all three of its forms: solid, liquid, and gas. That substance is water (H_2O), and its solid form, ice, has a number of important properties.

At standard atmospheric pressure (1.013 bar), water will solidify into ice at 32°F (0°C). Ice is crystalline, and solidification is a change of state from the random molecular movement of a liquid into a structure in which the molecules are held at fixed distances in a lattice (see CRYSTALS AND CRYSTALLOGRAPHY). As water freezes, the kinetic energy of the molecules is withdrawn from it as latent heat (see LATENT HEAT), and its temperature remains at 32°F (0°C) until all this heat has been given up. When ice melts, the reverse process takes place, and the water produced by the melting remains similarly at 32°F (0°C) until all the ice has melted.

The relative density of ice (compared with that of water, which is 0.9998 at 32°F) is 0.917—that is, the volume of ice is 9 percent greater than an equal mass of water at the same temperature. This property is what causes frozen water pipes to burst, and it is also responsible for the fact that ice floats in water, with one-tenth of its volume above the surface. This decrease in density upon solidifying is an unusual property but very important for life: oceans and lakes do not freeze from the bottom up; instead, the water below is protected from freezing by the insulating layer of ice at the surface.

Another result of the increase in volume upon freezing is that the freezing point becomes lower as pressure increases. In the school laboratory, this is often demonstrated by hanging a pair of heavy weights on a thin wire laid over a block of ice: under the pressure of the wire, the ice melts, and the wire gradually sinks through it to emerge at the bottom of the block. Ice skaters make use of this phenomenon; their thin blades produce water to lubricate their movement. The slipperiness of icy roads is due to the same cause. Glaciers flow for the same reason; their weight melts the ice in contact with the rocks beneath. Heavy objects left accidentally on ice fields for long periods of time have been lost forever because they sank into the depths.

CORE FACTS

- Ice is the solid state of water; it forms hexagonal crystals.
- At standard atmospheric pressure, ice is formed at 32°F (0°C).
- The freezing point of water is lowered as the pressure applied to it increases.
- The density of ice is less than that of water at the same temperature.
- The structure of ice crystals is due to hydrogen bonding between the water molecules.

The crystal state

Snowflakes are single crystals of ice, and their shape reveals the hexagonal symmetry of the crystal. However, depending upon the conditions in which they form, snow and frost crystals can be many different shapes—it is often said that no two snowflakes are the same. They can be simple stars, intricate branching stars, prisms, needles, hollow columns, or combinations of all these.

Snow and frost are formed directly from water vapor in the atmosphere, but ice formed from liquid water, including hail, is a compact mass of many different crystals. The crystalline nature is not, however, usually apparent. It can be revealed by focusing the Sun's rays with a magnifying glass on the interior of a piece of ice. This causes the ice to melt at the point where the rays are focused, causing a snowflake-shaped cavity, filled with liquid water, to form. This is known as a Tyndall flower, after Irish physicist John Tyndall (1820–1893), who first described it.

The ice crystal is built up from molecules of water linked by hydrogen bonds (see CHEMICAL BONDS). These bonds form because water is polar. In a water molecule, the single electron of each hydrogen atom is attracted toward the oxygen atom, so that the oxygen end of the molecule becomes slightly negatively charged and the hydrogen end becomes positively charged. The hydrogens will then be attracted to the negative oxygens of neighboring molecules.

This bonding causes the formation of complex molecules in liquid water, but the kinetic energy of the molecules is sufficient to keep breaking the bonds. As ice forms, however, the molecules become fixed in the crystal lattice. The length of the hydrogen-oxygen bond (the distance between the hydrogen atom and the neighboring oxygen) is 1.76 times the length of the hydrogen-oxygen bond inside the individual water molecule. Each water molecule becomes bonded to four other water molecules at the corners of a tetrahedron around it. The tetrahedral structure of ice is very similar to that of diamond (see CARBON) but not exactly the same. The formation of ice results in water molecules being joined together in layers, which can be joined to other layers above or below in any of six different arrangements.

Physical properties of ice

No ice crystal has a completely ordered structure. Here and there in the crystal there are water ions: the hydronium ion (H_3O^+) and the related hydroxyl ion (OH^-). Similar to the free electrons in a metal crystal (see METALS), these ions make ice conductive of electricity.

The crystal structure also affects the way in which ice behaves under shear stress. Under a strong enough "sideways" force, each layer can slide in relation to those above or below it, breaking bonds and forming new ones, so that the crystal shape is permanently deformed.

Snow will reflect up to 90 percent of the light falling on it, but glacier ice is much less reflective.

Both snow and ice will absorb almost all infrared radiation, but, when dry, they are relatively transparent to radio and microwaves. Radar can therefore be used to measure the depth of dry glaciers, even when they are a few miles thick.

S. FENNELL

The crystalline structure of ice, known as the Tyndall flower, was first described by Irish physicist John Tyndall.

See also: CHEMICAL BONDS; CRYSTALS AND CRYSTALLOGRAPHY; WATER.

Further reading:

Green, B. *Water, Ice, and Stone: Science and Memory on the Antarctic Lakes.* New York: Harmony Books, 1995.
McMurray, J. *Chemistry.* Englewood Cliffs, New Jersey: Prentice-Hall, 1995.

FREEZING FOOD

Because water increases in volume as it solidifies, freezing certain foods presents problems. The cell walls of soft fruits, for example, are often ruptured by the expanded ice, causing them to become mushy when thawed.

In certain conditions, when there is only a small amount of water vapor in the atmosphere, ice and snow will sublime—that is, go directly from solid to vapor without first melting to liquid. This property is used when foods are freeze-dried. The food is first frozen and then exposed to a high vacuum. The water content is removed as vapor, only about 2 percent remaining in the dry food.

SCIENCE AND SOCIETY

ICE AGES

Ice ages are cold climate spells during which glacial ice spreads over regions that are not normally covered in ice

The ice age landscape would have been dominated by large ice sheets covering the affected areas of Earth's surface, as shown in the artist's representation above.

CONNECTIONS

● Astronomic factors, namely the movements of **EARTH** in relation to the **SUN**, are believed to have played a major part in the timing of the ice ages.

● Glacial deposits and **LANDFORMS**, produced during the ice ages, have helped in dating **GEOLOGIC** time periods.

There is geological evidence for at least seven ice ages during Earth's history, each lasting millions of years. There is evidence of major Precambrian ice ages about 2.7 billion, 2.3 billion, 900 million, and 650 million years ago (see GEOLOGIC TIMESCALE). Later ice ages occurred in Phanerozoic time at the end of the Ordovician period (from 464 to 438 million years ago), the end of the Carboniferous period (from 320 to 286 million years ago), and the end of the Quaternary period (which started 1.6 million years ago), the last of these being known as the Quaternary ice age. During these ice ages there seems to have been cyclic fluctuations, with many

CORE FACTS

■ Several ice ages have occurred during the history of Earth, at what appear to be long and irregular intervals.

■ The causes of ice ages seem to be the movement of the continents by plate tectonics, which changes the circulation of ocean currents and weather patterns, and the position of the continents relative to the poles.

■ We are now in a warm, interglacial interval within the Quaternary ice age. This has lasted about 10,000 years, but a new glacial interval is likely to occur in the next 20,000 to 40,000 years.

repetitions of cold glacial and warm interglacial phases. Many geologists believe that we are currently living in a warm interglacial phase of the Quaternary ice age and that temperatures will fall and ice sheets will return in another 20,000 to 40,000 years.

Evidence of ice ages

Ice ages produce large systems of glaciers and extensive ice sheets similar to, but larger than, those currently covering much of Antarctica and Greenland (see ANTARCTICA). As snow accumulates on these ice sheets and glaciers, the cumulative weight causes the solid but semiplastic ice to deform with gradual creep and flow. It moves slowly outward and downward, scraping and eroding the underlying rocks.

At the end of an ice age, when the ice retreats, deep grooves and striations (smaller grooves) can be seen in the underlying rocks, and several distinctive deposits are left as evidence of ice action. These include till, often called boulder clay, which consists of abraded boulders and pebbles in a matrix of fine-grained, bluish silt; erratics, which are boulders that may weigh many tons and may be carried hundreds of miles by the ice before being dumped; and outwash sands and gravels arranged as undulating knolls, hummocks, sheets, or ridges. Sometimes these glacial sediments are laid down as stratified deposits. They may form fluvial terrace deposits on valley sides and beach terrace deposits along the former shorelines of ice and moraine-dammed lakes.

Other evidence of ice ages includes stratified varve sediments (layers of sediment deposited in a single year) in lakes and glacial, marine sediments on the ocean floor. Such sediments include drop stones, which were originally caught up in glacial ice and floated out with the ice into lakes and the sea. As the icebergs melted, they released the boulders—often far from the shore and their point of origin—and dropped them into the soft sediments below with their characteristic impact impressions.

Most striking of all the evidence of glaciations is the modification of the landscape by massive abrasion of mountains and valleys, and the overall wearing down of great areas of the continents by ice sheets armed with eroded rock fragments.

The weight of an ice sheet causes Earth's crust to subside. After the ice retreats, the crust slowly rises again. Even today, Scandinavia and the Hudson Bay region of Canada are rising nearly ⅖ in (1 cm) a year because the crust in these regions is still adjusting to the removal of an ice load about 12,000 years ago. One visible result is the presence of benchlike structures, called raised marine terraces, and beach deposits (see COASTS), which mark earlier shorelines with all their characteristic erosive and depositional features around the coasts of eastern Canada, Britain, and Scandinavia. These originated on the seafloor, and although the sea level has risen with the melting of the ice sheets, the crust in these regions has risen even faster so that what were once wave-cut platforms are now well above sea level.

Precambrian ice ages

Tillites (deposits left behind after the retreat of glaciers and ice sheets) have been recorded from hundreds of Precambrian localities around the globe. Often these are associated with striated (grooved) bedrock, varved sediments, and drop stones. Late Archean tillite from the Bruce Formation in Ontario, Canada, has been found to be 2.7 billion years old. Proterozoic tillites (tillites from 2.5 billion years ago to 570 million years ago) in the Gowganda Formation in Ontario, and other tillites of similar age in Quebec, Michigan, and Wyoming, are 2.3 billion years old. Tillites of about the same age occur on other continents, but the dating is not precise enough to determine whether they are all part of a single global glaciation or whether there were a number of glacial events. Widespread late Proterozoic glaciation occurred about 900 million years ago and again about 700 to 650 million years ago.

Phanerozoic ice ages

The evidence of successively younger glacial events is better preserved and often more widespread. A

As snow accumulated during the ice ages, it would have become compressed under its own weight to form glaciers, much as it does in polar regions today.

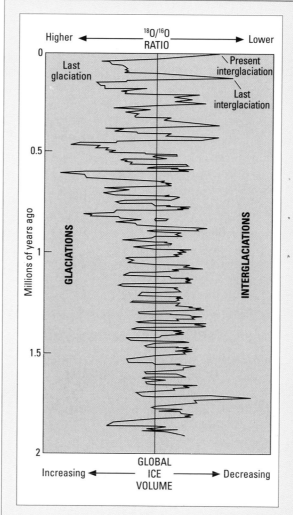

DEEP SEA DRILLING PROJECT

Ocean-drilling projects, such as the Deep Sea Drilling Project carried out by the drilling ship *Glomar Challenger* from 1968 until 1983, and the succeeding Ocean Drilling Project carried out by the drilling ship *JOIDES Resolution* from 1983 to the present, have opened up a new world for glaciation studies. Deep-sea sediments provide excellent evidence of glacial and interglacial cycles. Variations of the oxygen isotope ratios in small planktonic shells (that is, the shells of organisms inhabiting the surface layer of water) show repeated shifts in the oxygen isotope composition. These isotope variations represent changes in global ice volume. When water evaporates from the oceans and precipitates on land to form ice sheets, water containing the light isotope oxygen-16 is evaporated more readily than water containing the heavier isotope oxygen-18. As a result, the ice sheets become enriched with the lighter isotope, while the oceans become enriched with the heavier one. Isotope curves derived from deep-sea sediments provide a continuous record of changing ice volume on Earth's surface. This method, coupled with radioactive dating to provide a time frame, is now well-established as the standard record of glacial and interglacial cycles during the past two million years.

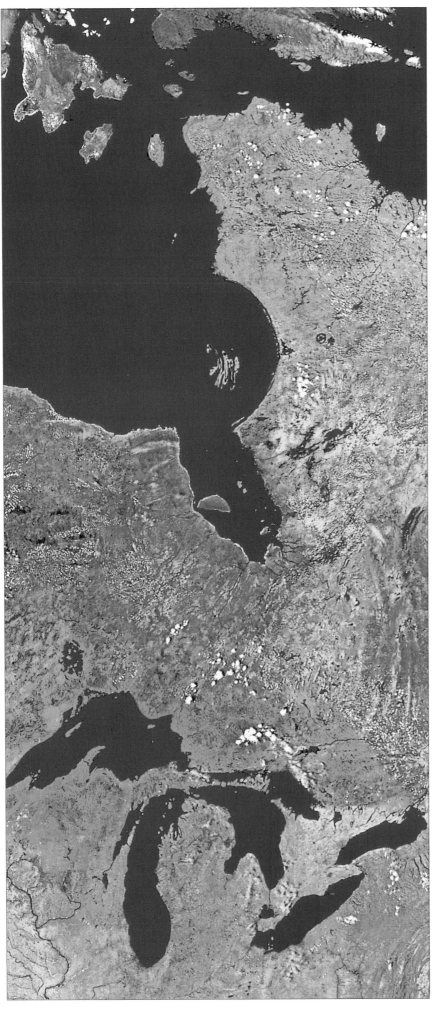

global drop in sea level toward the end of the Ordovician period, 438 million years ago, is considered to be the result of glaciation. Tillites and striations have been found in North and South Africa, South America, and western Europe. The drop in sea level resulted in a considerable reduction in the area of shallow seas. It was this, together with the increase in the oxygen content of surface sea water because of cooler temperatures, that caused a major extinction of marine animals, known as the End-Ordovician extinction event (see EXTINCTIONS).

Another, even greater, glaciation occurred toward the end of the Carboniferous period and extended into the Permian period. At that time, the continents of the Southern Hemisphere, grouped together as the supercontinent we now call Gondwana (see GONDWANA), moved over the South Pole, resulting in extensive glaciation that affected different areas of Africa, South America, India, and Australia and producing changes in sea level. This is reflected in sedimentation patterns of rock types such as tillites and drop stones (an indication of glacial transportation and deposition). It is also reflected in the significant global decrease in faunal diversity (that is, the range of organisms that existed globally). There is some recent evidence, however, that other reasons for the extinctions may be just as important—a decrease in inland seas as well as the lowering of sea level could have caused a reduction in habitat for shallow marine organisms.

The most extreme impact of this glaciation was in causing ice sheets to extend well away from the South Pole and toward the equator, with evidence of cool temperatures worldwide.

The Quaternary ice age

About 1.6 million years ago, the Quaternary ice age started. It was characterized by a number of advances and retreats of ice sheets. The ice extended in Europe to cover all of Scandinavia, most of Britain, and much of northern Germany, northern Poland, and northern Russia. There were also extensive separate ice sheets covering the mountains of the Alps, Pyrenees, Urals, and Himalayas. In North America, the ice covered all of Canada and New England. It also extended down into all the northern states of the United States and was particularly extensive south of the Great Lakes, into Minnesota, Wisconsin, Illinois, Michigan, Indiana, Ohio, and northern New York. As these great southern lobes of continental ice melted, they formed vast proglacial lakes, the remnants of which are now to be seen in the Great Lakes. Greenland and Iceland were completely covered, and the ice sheet that covered so much of North America was continuous across the North Atlantic with the ice sheet that covered northern Europe. Alaska and eastern Siberia were relatively free of ice, so it is evident that the ice sheet was centered more over the North Atlantic and

During the Quaternary ice age, northern ice sheets extended as far south as the Great Lakes, which were formed as the ice sheet melted and retreated.

CHANGING VIEWS OF THE ICE AGES

Some 18th-century scientists thought that large boulders and other glacial deposits were simply material transported by the biblical flood. In 1830, British geologist Charles Lyell (1797–1875) suggested that the boulders had been carried by large icebergs that floated in the water of the flood. Sediments, including the large boulders he visualized had drifted with the flood, were called drift, a term still used for these deposits.

The glacial theory, in which drift is explained as deposits from large glaciers, was developed in the European Alps during the late 18th century and early 19th century. In the Alps, glacial features can be observed adjacent to existing glaciers, including large transported boulders resting upon striated surfaces and moraine at the edge of the ice. In 1787, Swiss lawyer B. F. Kuhn, and Scottish geologists James Hutton (1726–1797) in 1795 and John Playfair (1748–1819) in 1802, reached the conclusion that the widespread boulders and moraine they observed great distances from the present glaciers were transported to their present location by former, more extensive glaciers. In 1832, German scientist A. Bernhardi described stones and boulders of Scandinavian origin found in Germany. Bernhardi suggested that a polar ice cap had reached as far south as southern Germany. Another noted exponent of the glacial theory was Swiss-born U.S. naturalist Louis Agassiz (1807–1873), who in 1837 proposed that ice sheets had covered much of northern Europe and North America.

The first scientists to relate the raised and tilted marine shorelines of northern Europe to glaciation were French mathematician J. Ademar in 1842 and Scottish scientist C. Maclaren, also in 1842. They related the present high-lying shorelines to changes in sea level caused by fluctuations in the volume of ice age glaciers. In 1865, Scottish scientist T. Jamieson attributed the raised terraces along the northern European shorelines to the unloading of the heavy ice sheets and the corresponding isostatic rise of Earth's crust. In 1863, Scottish geologist Sir Archibald Geikie (1835–1924) presented evidence of warm periods separating glacial advances in Scotland, based upon warm-climate fossils in sediments sandwiched between glacial beds. By the early 20th century, many scientists recognized evidence of four or more major Pleistocene glaciations separated by warm interglacial periods.

One of the theories proposed to explain alternating long-term and short-term climatic cycles is the astronomical theory. Several scientists were responsible for this theory, including Scottish physicist John Croll in 1864, and Serbian engineer and astronomer Milutin Milankovitch in 1911 (see also the box on page 597). The astronomic factors, also called the Milankovitch factors, include the eccentricity of Earth's orbit; the obliquity; or tilt, of Earth's axis; and the precession, resulting from the wobbling of Earth's axis and changes in eccentricity.

During the 1940s and 1950s, U.S. chemist Harold Urey (1893–1981) and Samuel Epstein developed the use of the ratio of oxygen-18 to oxygen-16 isotopes in shells to measure past changes in ocean temperature. This method and other evidence from microscopic planktonic sea shells preserved in oceanic sediments was used during the 1970s and 1980s to measure the ice age temperature changes recorded in deep-sea drilling samples. One of the leading workers in this field has been English geologist Nicholas Shackleton, who has helped relate oceanic temperature changes to glacial and interglacial fluctuations on land and has significantly refined the dating of temperature fluctuations during the past three million years.

DISCOVERERS

---

that the North Pacific was relatively ice-free. At times of major ice advance, so much water was "locked away" in the ice sheets that global sea level fell by more than 490 ft (150 m) and large stretches of the continental shelves emerged as dry land.

The ice sheets eroded great amounts of sediments and other rocks and deposited the eroded materials as glacial sands, till, and other debris over much of the area covered by the ice and also over outwash plains (flat plains that comprise mostly deposits of rivers that issued from the front of ice sheets) in the peripheral regions.

There is field evidence on land of six major advances of the ice sheets, with intervening interglacial warm intervals. This understanding is based on the deposits left by successive advances of the ice in western Europe and North America. In Europe, the names of the ice advances come from areas in the glacial valleys north of the Alps. These advances are called the Biber, Donau, Günz, Mindel, Riss, and Würm. Each time the ice returned after an interglacial period, it removed much of the drift and moraine left by the previous glaciation, so that the earlier advances were increasingly more difficult to identify. In North America, this effect makes it hard to detect the earlier advances.

In North America, the four most recent advances, named after states representing the farthest encroachments, are the Nebraskan, Kansan, Illinoian, and Wisconsin glaciations. There is excellent evidence, obtained from oxygen isotope analysis (see the box on page 593) of calcite deposits in a fissure in Nevada, that the six glaciations known from the Alps did occur in North America. The measures can be matched to the standard sequence of consecutive isotope stages (see ISOTOPES). They show that the same six glaciations are also well-represented by moraines (accumulations of material that has been transported or deposited by ice) preserved in Britain, the Netherlands, and elsewhere in northern Europe.

Because of the difficulty in recognizing the earlier field deposits, there is a tendency today to abandon the names of at least the oldest, more problematic glaciations and to use the marine isotope stage numbers instead. This isotope evidence, together with some field evidence, leaves little doubt that there must have been many more than six glaciations, and that these extend back over two million years into the late Tertiary period. The isotope evidence from temperature-sensitive marine fossils suggests that mean summer temperatures during glacial times were about 60°F (15°C) below those of the interglacial times.

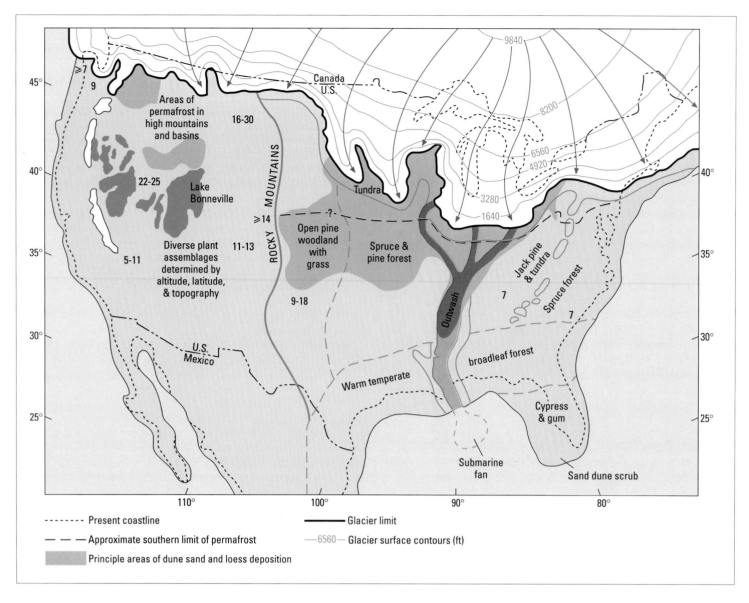

Legend:
------- Present coastline
— — — Approximate southern limit of permafrost
[shaded] Principle areas of dune sand and loess deposition
——— Glacier limit
—6560— Glacier surface contours (ft)

The diagram above shows a reconstruction of central North America during the last glacial maximum 18,000 years ago. The figures within the map show the amount the temperature dropped (°F).

Causes of ice ages

We now have clear evidence that cyclic climatic changes have been occurring for the last several million years and probably throughout Earth's history. They are caused by astronomic Milankovitch factors (see the box on page 597), which result in fairly predictable variations in the amount of solar radiation Earth receives and can partly explain the background climatic fluctuations that we see in the alternation of glacial and interglacial cycles during the Quaternary ice age. However, these are not great enough to explain the occurrence of the major Precambrian and Phanerozoic ice ages, nor do they explain the very irregular occurrence of those ice ages. They also do not account for the occurrence of the Quaternary ice age, nor for the absence of ice ages for 280 million years after the Permo-Carboniferous ice age.

The cause of ice ages is most likely to be the complex interplay of slow geographic and geologic changes ensuing from plate tectonics (see PLATE TECTONICS). These changes result in the movement of continents as they are carried along on the plates of the lithosphere; the uplift and folding of continents and the creation of mountains when these plates col-

lide; the opening and closing of ocean basins; the alteration of ocean circulation; the growth or breakup of supercontinents; and the repositioning of continents near the equator or the poles.

Ocean currents are important in spreading Earth's surface heat. Disruptions in this circulation may cause local heating and cooling. Diversion of wind systems by mountain chains may also cause regional climatic changes. Large landmasses at or near the poles seem to increase the chances of a buildup of large ice sheets. These climatic factors may interact in sensitive ways and are also likely to be affected by atmospheric carbon dioxide levels, which also may be altered by the growth in ice sheets once an ice age begins.

The ice age to come

The most recent ice age is the Quaternary ice age. The ice began to spread during the late part of the Pliocene, over two million years ago. A more general drop in temperatures started much earlier, about 30 million years ago during the Oligocene epoch. There was a steady fall in temperature through the Oligocene, Miocene, and Pliocene epochs, and ice sheets began to grow markedly during a cold cycle about 2.5 million years ago. Glacial and interglacial cycles have occurred since that time, but with a trend toward colder glacial phases. The most recent of these glaciations began about 115,000 years ago; it is called the Wisconsin glaciation in North America and the Würm glaciation in central Europe. Immediately prior to that glaciation, hippopotamuses and lions lived in Europe and across Asia in places with a semitropical climate. Indeed, the presence of this interglacial period was felt worldwide. Some warmer periods occurred during the Wisconsin glaciation, but then a long period with a predominantly cold climate ensued, resulting in ice sheets that covered much of northeast North America and northwest Europe. The temperatures began to rise again about 12,000 years ago, and present-day temperatures were reached about 8000 years ago, during the Holocene epoch.

In view of the background cyclicity that has characterized global temperatures as far back as reliable isotope records extend, the present, relatively warm climate can be expected to continue for several thousand years, perhaps even for another 20,000 years. There is good reason, however, to consider the present Holocene epoch as something of a misnomer. It is hardly a distinct epoch but merely an interglacial cycle within the Quaternary ice age.

The standard method for predicting future climates is extrapolation. Scientists use past climate trends to estimate what is likely to happen in the future. Using the Croll-Milankovitch astronomical variations as a guide, scientists believe the next glacial cycle, or next ice age, should be expected to begin within the next 20,000 to 30,000 years.

N. SAVAGE

See also: GEOLOGIC TIMESCALE; GLACIERS; ICE; QUATERNARY PERIOD.

Further reading:

Andersen, B. G. and Borns, H. W., Jr. *The Ice Age World.* Norway: Scandinavian University Press, 1994.
Ferguson, S. A. *Glaciers of North America: A Field Guide.* Golden, Colorado: Fulcrum Publishing, 1992.
George, M. *Glaciers.* Mankato, Minnesota: Creative Education, 1991.
Hamblin, W. K. *Glacial Systems.* New York: Macmillan, 1992.
Hamblin, W. K. and Christiansen, E. H. *Earth's Dynamic Systems.* 7th edition. Englewood Cliffs, New Jersey: Prentice-Hall, 1995.
Hambrey, M. J. *Glacial Environments.* Vancouver: UBC Press, 1994.
Monroe, J. S. *Physical Geology: Exploring the Earth.* St. Paul: West Publishing Company, 1992.
Montgomery, C. M. *Physical Geology.* 3rd edition. Dubuque: William C. Brown, 1993.
Skinner, B. J. and Porter, S. C. *Glaciers and Glaciation in the Dynamic Earth.* New York: John Wiley & Sons, 1995.

THE MILANKOVITCH ASTRONOMIC FACTORS

Several factors related to positions of the Sun and Earth affect the amount and distribution of solar radiation. These were described by several researchers, including Serbian engineer and astronomer Milutin Milankovitch (see also the box on page 595), and include three main effects: eccentricity, obliquity, and precession. Earth circles the Sun each year in an eccentric orbit (that is, not precisely circular), and the shape of the eccentric orbit slowly changes. In addition, Earth wobbles on its axis, causing changes in the date at which Earth is closest to the Sun, called perihelion. The combined motions of the wobble and the change in eccentricity cause a progressive shift, or precession, of the spring and autumn equinoxes, with each cycle lasting about 23,000 years. Another factor is variation in the tilt of Earth, or obliquity. This changes from a minimum of about 22 degrees to a maximum of about 25 degrees. It is currently at 23.5 degrees. The angle of tilt affects the difference in temperature from summer to winter and can affect the amount of accumulated winter ice that melts during the succeeding summer.

All these factors have to be combined to calculate the changes in radiant heat from the Sun that affect different parts of Earth. They probably account for the climatic fluctuations that cause the alternation of glacial and interglacial intervals during an ice age and probably produce comparable climatic fluctuations during the much longer periods when Earth is not experiencing an ice age.

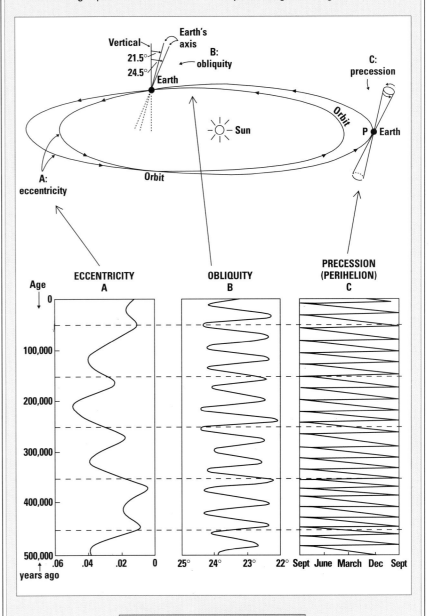

A CLOSER LOOK

IGNEOUS ROCKS

Igneous rocks are rocks formed directly by the cooling and solidification of magma

Columnar jointing of basalts is quite common and produces the structures shown above on the Isle of Staffa, Scotland.

CONNECTIONS

- Igneous rocks are often very resistant to **EROSION** and may form prominent **LANDFORMS**.

- Igneous **ROCKS** are formed from lava that flows from **VOLCANOES** and cools and solidifies.

Igneous rocks are formed from the molten rock, or magma, that comes from deep below Earth's surface, and they tell the story of the formation of Earth's crust. It is estimated that about 95 percent of the crust is composed of either igneous rock or metamorphic rock formed from igneous rock by heat and pressure (see METAMORPHIC ROCKS). Igneous rocks are not underfoot everywhere, however, because much of the crust is topped by a thin veneer of sedimentary rock, made from weathered rocks of all types (see SEDIMENTARY ROCKS).

Igneous rocks get their name from the Latin word *ignis*, meaning "fire." Basalt, the most common igneous rock, is formed when lava flowing from a volcano cools and solidifies. Other forms of igneous rock, such as the tuff that entombed the victims of the Mt. Vesuvius eruption of 79 C.E. at Pompeii, are also of volcanic origin. Basalt and tuff are extrusive igneous rocks, which are rocks that have formed by the solidification of magma poured out onto Earth's surface. But magma can also cool and solidify deep

beneath the surface, forming intrusive igneous rocks such as granite. As tectonic forces and erosion slowly but continuously rework the face of Earth, rocks can change from one form to another. Any rock can be melted to become a new igneous rock.

CORE FACTS

- Igneous rocks are formed when magma from Earth's interior cools.
- Metamorphic rocks are formed from igneous rocks by heat and pressure. About 95 percent of Earth's crust is composed of either igneous or metamorphic rocks.
- Igneous rocks can be extrusive, formed by eruption of lava onto Earth's surface, or intrusive, cooling and solidifying deep below the surface.
- Basalt is a typical extrusive rock, and granite is an intrusive rock.
- The classification of igneous rocks is based on their composition and texture.

Igneous rock texture

Because igneous rocks start out as molten material, their texture, when cooled, is generally characterized by interlocking crystals. The more slowly a magma cools, the larger the crystals can grow. Basalt has an aphanitic texture, composed of interlocking crystals too small to be seen by the naked eye. Phaneritic rocks such as granite are characterized by visible, interlocking crystals several millimeters in size. Phaneritic rocks are formed when magma cools slowly underground; they are found at Earth's surface only after mountains have formed and subsequently eroded.

Some kinds of igneous rock (porphyrites) can have a porphyritic texture, characterized by large, well-formed crystals in a fine-grained groundmass (the material making up the main body of the rock). The large crystals, called phenocrysts, are formed during an initial stage of slow cooling. When the magma is then brought to the surface, it finishes cooling quickly, forming, for example, a basalt porphyry.

Rocks composed of volcanic fragments are called pyroclastic rocks. Pyroclastics are fragments hurled into the air by explosive eruptions. The fragments range in size from ash (microscopic glassy particles) to large volcanic bombs, hand-sized or larger.

An accumulation of loose pyroclastics is called tephra, and cemented tephra is called tuff. An ash-flow tuff may look like an aphanitic rock, but under a microscope it becomes clear that it is composed of broken crystal fragments of a glass-like nature rather than interlocking crystals. Glassy rocks, such as obsidian and pumice, have no crystalline structure at all. They are formed when molten lava is cooled instantaneously or quenched by air or water. Basalt flows often have a glassy surface.

Composition of igneous rocks

Magma, the source of all igneous rock, is a very hot fluid mass made up of the most abundant elements in

IGNEOUS LANDSCAPES

The Sierra Nevada batholith, formed during the age of dinosaurs, contains some of the most closely examined rocks in the world. While rock climbers in Yosemite Valley may be searching for handholds on the smooth faces of Half Dome and El Capitan, geologists study the granite rocks for clues to their origins. The entire batholith (a large intrusive mass of igneous rock), extending about 400 miles (650 km) north-south and 60 miles (100 km) east-west, is actually made up of about 10 smaller sub-batholiths, with distinct compositions and textures.

The batholith is composed of granite and, together with similar batholiths in Idaho and Baja California, marks a once active subduction zone that at one time stretched the length of the North and South American continents. The Sierra Nevada batholith was uplifted by tectonic forces, and erosion has removed most of the 6 miles (10 km) or so of overlying rock, leaving scattered roof pendants, remnants of the original roof of the magma chamber. During the Pleistocene epoch, beginning nearly 1.6 million years ago, glaciers carved out Yosemite Valley, leaving a trail of striations, or scratches, where rocks embedded in the glacial ice scraped across the bedrock.

A CLOSER LOOK

The light micrograph below shows a thin section of a sample of granite. The large crystals seen throughout represent inclusions of a preexisting rock.

The volcanic plug at Cape Schrank, Australia, is the remnant of the feed channel of a volcano that has been filled with solidified lava and pyroclastic material. The surrounding part of the volcano has subsequently eroded to reveal the central plug, which is more resistant to erosion.

Earth's crust: oxygen, silicon, aluminum, iron, magnesium, calcium, sodium, and potassium. The viscosity, or stiffness, of a magma is determined by its content of silica (silicon dioxide, SiO_2) and water vapor. Silica in a magma increases its viscosity, whereas water decreases its viscosity.

There are three main types of magma: basaltic, andesitic, and rhyolitic, named after the corresponding extrusive rocks they form.

Basaltic magmas, the most abundant of the three types, are relatively low in silica (about 50 percent) and rich in iron and magnesium. They are heavy and black and flow easily.

Andesitic magmas, second in abundance to basaltic, are formed in subduction zones—where the surface of Earth under the ocean moves down into the interior—such as the ones now surrounding the Pacific Ocean.

Rhyolitic magmas are highest in silica (about 70 percent) and are very viscous, so they flow slowly. Rhyolitic magmas can hold more gas in solution than

basaltic magmas because of their relatively low temperatures. Rhyolitic magmas rich in dissolved gases produce violently explosive eruptions.

Common forms of extrusive igneous rock

Basalts that originate from surface lava flows are the most common type of extrusive igneous rocks. Underwater basaltic eruptions at spreading centers in midocean ridges form bulbous pillow basalts on the ocean floor. The volume of these flows can be immense: the entire oceanic crust was formed from these spreading centers.

Huge quantities of basaltic magmas have also erupted on the continents as flood basalts. Because basaltic lava flows easily, eruptions on land typically form thin sheets. A single flow may be hundreds of miles across but is usually only about 50 ft (15 m) thick. The Columbia Plateau flood basalts in Oregon and Washington began pouring from fissures in the crust 17 million years ago. Over a period of 3 million years, flows followed one another to cover some 50,000 sq miles (200,000 km^2) to a depth of 5000 ft (1500 m). Flood basalts are often interlayered with sedimentary rocks.

The texture of a basalt flow depends on its viscosity and the steepness of the terrain. Lower viscosity magmas, or fast flowing magmas, form smooth or ropy lava called pahoehoe lava, whereas magmas with higher viscosity, or slower flowing magmas, form jagged, blocky lava, called aa lava. (These names come from Hawaii.) Vesicular lava contains spherical or flattened holes called vesicles, formed when gas in the magma forms bubbles near the surface.

Igneous rocks shrink as they solidify, often resulting in large-scale cracking, called jointing. Thick basalt flows typically fracture into a pattern of hexagonal-shaped columns, from about half a yard to a few yards in diameter. At the Devils Postpile in California, glaciers scraped over the top of such a columnar-jointed basalt flow. The tops of the hexagonal columns were polished so smooth that the surface resembles a parquet floor.

Andesitic and rhyolitic magmas are very viscous, so they tend to form stubby domes, plugs, or steep-sided cones rather than flows. Rhyolitic magmas rich in volatile gases can produce ash-flow tuffs. Ash flows are created when ash and rock fragments are ejected and suspended by hot, expanding gases. The dense, hot mixture can surge downhill at up to 125 mph (200 km/h), incinerating everything in its path. Welded tuff, or ignimbrite, is formed when an ash flow comes to rest while still very hot and the ash and rock fragments become fused together.

Intrusive rock structures

Our knowledge of intrusive rock forms is limited by the fact that we live on Earth's surface. Since they cannot directly observe their formation, geologists must rely on erosion to expose these structures for study. The most common forms of intrusive igneous rocks are large bodies called plutons, such as batholiths, and sheetlike structures, such as dikes and sills.

Batholiths, named from the Greek word *bathos*, meaning "depth," are coarse-grained rock bodies extending over 40 sq miles (100 km²) or more. They originate from hundreds of miles below Earth's surface. Batholiths are formed in active mountain-building regions, such as the Andes in Peru, where the crust below the ocean is being subducted into Earth's interior and remelted there. Molten, tear-shaped bodies of andesitic or rhyolitic magma, called diapirs, are formed by partial melting of the oceanic crust and then rise through Earth's crust.

Batholiths are generally shaped something like a molar tooth, flat on top, with a base about 4 to 6 miles (6 to 10 km) deep and roots extending down over 10 miles (15 km) into the crust. To understand how batholiths can displace huge volumes of rock as they rise through the crust, geologists look for clues in the batholith and in the surrounding ("country") rocks. Irregular chunks of the country rock, called xenoliths, are sometimes incorporated in the batholith. These are thought to be formed when the roof of the magma chamber fractures and collapses, making room for the magma to rise.

Often the margins of batholiths have smaller crystals than their interiors; this is because the outer areas cool more quickly than the interior. Surrounding rocks may be melted and deformed or just pushed aside with no signs of heating. Stocks are like batholiths, only smaller and often cylindrical.

Dikes and sills are sheetlike bodies, from a few inches to several yards thick. Dikes cut across layers of intruded rock, and sills are formed parallel to existing rock layers.

In regions where the continental crust is extending, dikes can occur in groups of tens or hundreds of parallel sheets called swarms. These form when magma rises from the mantle through underground fractures. Swarms of dikes found in Ireland, Greenland, and New England are all associated with the opening of the North Atlantic ocean basin. Dikes are also found in association with volcanoes.

Large sills sometimes form laccoliths. A laccolith is a sill formed between sedimentary layers close to Earth's surface. Laccoliths are flat on the bottom, parallel to existing rock structure, but arch upward on top. They are usually formed at depths of less than 2 miles (3 km), where the burden of overlying sediments is light enough to lift up.

J. FEDERIUK

See also: LANDFORMS; MAGMA; METAMORPHIC ROCKS; ROCKS; SEDIMENTARY ROCKS; VOLCANOES.

Further reading:
Blair, G. *The Rockhound's Guide to Arizona.* Helena, Montana: Falcon Press, 1993.
Cunningham, C. G. *et al. Ages of Selected Intrusive Rocks and Associated Ore Deposits in the Colorado Mineral Belt.* Washington: U. S. G. P. O., 1994.
Hamblin, W. K. *Introduction to Physical Geology.* 2nd edition. New York: Maxwell Macmillan International, 1994.

HOW ARE IGNEOUS ROCKS CLASSIFIED?

Igneous rocks are classified on the basis of their mineral constituents and on their texture. The composition of the rock provides clues about where and how the magma originated, and the texture reveals how the rock cooled. Igneous rocks are composed of silicate minerals; that is, minerals containing the two most abundant elements in Earth's crust: oxygen and silicon. Other common elements in the silicate minerals are aluminum, iron, magnesium, calcium, sodium, and potassium. Minerals rich in silica, such as quartz and the feldspars (orthoclase and plagioclase), are called felsitic. These minerals are generally light in color, light in weight, and have relatively low melting temperatures.

The mafic minerals are high in magnesium and iron and low in silica. They are generally heavy and dark. Common mafic minerals include biotite, amphibole, pyroxene, and olivine.

The terms *felsitic* and *mafic* are also applied to rocks composed primarily of either felsitic or mafic minerals. Of the coarse-grained intrusive rocks, granite is felsitic, diorite is intermediate, and gabbro is mafic. Peridotite, an intrusive rock composed of olivine and pyroxene, contains even more iron and magnesium than gabbro does and is therefore called an ultramafic (or ultrabasic) rock. Earth's mantle (the part between the crust and the core) is largely composed of peridotite.

Basalt, the most common volcanic rock, is a mafic rock composed mainly of calcium plagioclase and pyroxene, with small amounts of olivine or amphibole. Basaltic magmas rich in dissolved gases will form vesicular basalt, or scoria. If the vesicles are later filled with minerals, the basalt is called amygdaloidal.

Granite is a light gray or pinkish rock composed mainly of quartz and feldspars. Its crystals are large enough to identify easily; they usually include shiny black flecks of biotite and needlelike crystals of hornblende. Granite is fairly light, with a specific gravity of 2.7 (basalt has a specific gravity of 3.2). The extrusive equivalent of granite is rhyolite.

Peridotite is a coarse-grained rock composed predominantly of the mineral olivine.

A CLOSER LOOK

INDIAN OCEAN

The Indian Ocean is Earth's third largest ocean

This image shows part of the Indian Ocean as seen from space. India can be seen at the top of the picture and the Maldive Islands to the left.

Third in size of Earth's oceans, the Indian Ocean has several unique features that make it different from the larger Pacific and Atlantic. For example, the Indian Ocean is landlocked on its northern boundary and, as a result, does not form part of the colder climates of the Northern Hemisphere. Perhaps even more unusual is the fact that many of the Indian Ocean's currents reverse themselves twice each year in response to the tropical wind patterns that give rise to the monsoon seasons in India and southern Asia.

The coastline surrounding the Indian Ocean stretches for 40,000 miles (64,000 km). It runs from India along the Bay of Bengal, down through the islands of Indonesia, along the west coast of Australia, past Antarctica, up to Africa, along the Arabian peninsula, across the Persian Gulf, and back to India. Within these boundaries is 28.4 million sq miles (73.4 million km²) of surface land area, or about one-fifth of Earth's total oceanic surface. With an average depth of about 12,700 ft (3850 m), the Indian Ocean contains 10.3 billion cubic ft (293 million m³) of seawater. A number of marginal seas and other stretches of water, such as the Arabian Sea, Strait of Malacca, Java Sea, Red Sea, Persian Gulf, and Timor Sea, are also considered as part of the Indian Ocean. In many places, subsurface mountains and plateaus rise out of the water to form islands such as Madagascar, Sri Lanka, Reunion, Mauritius, and the Seychelles.

Origin

The Indian Ocean is younger than the Atlantic and Pacific. In fact, geologic evidence indicates that it was formed entirely during the past 100 million years. Scientific expeditions, such as those conducted in the 1970s as part of Deep Ocean Drilling Project, confirmed earlier theories that the Indian Ocean owes its creation to seafloor spreading and plate tectonics that resulted from the breakup of Gondwana, one of the two ancient supercontinents produced by the first split of the even larger supercontinent Pangaea about 200 million years ago (see GONDWANA; PANGAEA).

Oceanic currents

The Indian Ocean's seasonal reversal of currents is the result of changes in wind patterns. In the winter months, the northeast monsoon results in strong winds that push currents in a westerly direction, making them travel to Africa and swing southward until reaching the equator. There, they turn back to the east as part of the equatorial countercurrent. This pattern is most well defined in February. During the summer, the opposite current pattern is created by the southwest monsoon, the monsoon that brings the wet season to India. The current slowly reverses itself until, by August, the dominant flow is northward along the African coast north of the equator and then eastward across the Arabian Sea. South of the equator, the prevailing currents remain west to east throughout the year.

Subsurface features

If drained of its water, the Indian Ocean would be a rugged landscape of deep basins and trenches separated by sheer-walled ridges, some as high as 2 miles (3.2 km). The most spectacular of these is the mid-

---

CORE FACTS

■ The Indian Ocean occupies about one-fifth of the planet's total oceanic surface and is the third largest ocean on Earth.

■ The Indian Ocean is younger than the Atlantic and Pacific Oceans. It was formed about 100 million years ago by the breakup of the ancient continent of Gondwana.

■ Currents in the northern parts of the Indian Ocean reverse their direction twice each year in response to the winds that create Asia's monsoon seasons.

ocean ridge. This steep-walled ridge runs southward from the Gulf of Aden and then branches like an inverted Y, one arm jutting to the southwest beneath the tip of Africa and the other curving gently away to the southeast, where it terminates between Australia and Antarctica. This massive ridge structure is seismically active, and the earthquakes along its length produce new oceanic crust at several locations. Although most of the margins of the ocean are free of seismic disturbances, a major exception occurs in the east, especially in the area of Sumatra. Here, the continued northward spreading of the ocean floor creates numerous earthquakes and volcanic eruptions. These are most concentrated near the Sunda Trench, where the Indian Ocean's greatest depth, 24,600 ft (7455 m), was recorded.

Minerals and other natural resources

Although 40 percent of the world's offshore petroleum production occurs in the waters of the Indian Ocean, the ocean's other resources remain largely untapped. For example, large quantities of natural gas are known to exist, but only about 1 percent of the world's total production of this resource comes from the Indian Ocean. Another valuable resource known to exist on the floor of the ocean are manganese nodules. These grapefruit-sized nodules form at depths of 2 to 3 miles (3.2 to 5 km) and typically contain manganese, iron, nickel, cobalt, and copper. At present, mining and processing costs hinder the large-scale operations designed to exploit this resource. Conflicting claims by the 35 nations that border the Indian Ocean have also severely limited efforts to develop the ocean's resources.

J. HUMPHREYS

See also: AUSTRALASIA; CURRENTS, OCEANIC; OCEANS AND OCEAN ABYSSES; PLATE TECTONICS.

Further reading:

Godfrey, J. S. *et al. The Role of the Indian Ocean in the Global Climate System.* College Station: Texas A & M University, 1995.
Wriggins, W. *Dynamics of Regional Politics: Four Systems on the Indian Ocean Rim.* New York: Columbia University Press, 1992.

ASEISMIC RIDGES

Oceanic ridges and other upland areas that are not volcanic today but may have been formed by the results of volcanic activity in the distant past are often called aseismic ridges. Several of these ridges are found in the Indian Ocean. The most well known is Ninety East Ridge. This is a high, seismically stable ridge that runs in an almost straight north-south line for 3100 miles (5000 km). The crest of this ridge is sometimes as much as 12,000 ft (3658 m) above the ocean floor. It is thought that Ninety East Ridge was formed as a result of the plate tectonic processes that led to India moving northward to its present position. Another aseismic feature of the Indian Ocean is the Chagos Laccadive Plateau. This is also thought to have been created as India drifted toward a collision with Asia. Located off the west coast of India, the plateau extends above the surface of the ocean at several places, most notably to form the Maldive and Laccadive Islands.

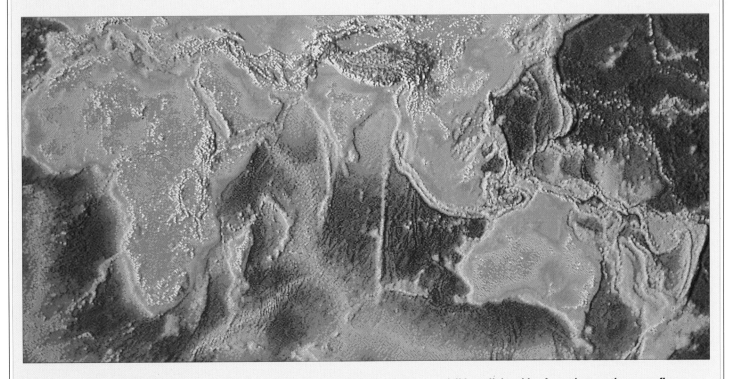

This computer map, centered on the Indian Ocean, shows the oceanic ridges of the area, visible as lighter blue formations on the ocean floor.

A CLOSER LOOK

INFRARED RADIATION

Infrared radiation is electromagnetic radiation with a wavelength slightly longer than that of visible light

This false-color infrared image shows the Bighorn Basin in Wyoming. The Bighorn River is seen running from top to bottom, near the right. This image was made by the Thermatic Mapper of a LANDSAT satellite in 1991.

CONNECTIONS

● Certain types of gas **LASERS** use **CARBON DIOXIDE** to create very high-powered infrared beams.

● Infrared **SPECTROSCOPY** is useful for studying the structure of **MATTER**.

Radiation from the Sun is our primary source of heat on Earth; without it, life on Earth would be impossible. When the Sun's radiation reaches the surface of Earth, it is composed of 49 percent infrared radiation, 43 percent visible light, 7 percent ultraviolet light, and less than 1 percent X rays, gamma rays, and radio waves.

With wavelengths ranging from 730 nm to 1 mm, infrared radiation is so called because it has a wavelength that is just longer than that of red light. It was discovered in 1800 by an English astronomer, William Herschel (1738–1822). He used a prism to bend sunlight onto mercury thermometers and observed that an invisible form of radiation, with a wavelength slightly longer than the red end of the visible spectrum (see ELECTROMAGNETIC SPECTRUM), caused the temperature indicated on his thermometers to rise. The phenomenon that Herschel observed is known as radiative heating. This type of heating differs from other types of heating in that it can be transmitted with no medium. This means infrared radiation can travel through the vacuum of space.

We are emitting infrared radiation all the time. Our bodies, like all warm or hot objects, radiate this electromagnetic energy continuously, though the amount emitted does vary with temperature, and some parts, such as the head, emit more than others. Our bodies also absorb this energy. For example, if you stand in front of a fire, you start to feel warm as your body begins to absorb infrared radiation. The amount of infrared radiation that an object emits depends on the temperature of the object, as well as on the type of material it is made from.

Not all materials absorb infrared radiation equally. Glass allows visible light to pass through but is opaque to infrared radiation; it traps infrared radiation instead of transmitting it. In our atmosphere, infrared radiation is absorbed mainly by water vapor and carbon dioxide, because these substances are particularly good absorbers. There are certain infrared wavelength ranges that are not absorbed.

CORE FACTS

■ Infrared radiation is a type of electromagnetic radiation with wavelengths of 730 nm to 1 mm.

■ Infrared radiation is absorbed and radiated by certain objects.

■ Wavelength ranges not absorbed by the atmosphere are called atmospheric windows.

■ Radiative heating from the Sun plays an important role in the heating of our atmosphere.

■ The infrared radiation emitted by a body can be recorded on a thermogram; thermograms have important applications in medicine.

■ Infrared radiation is also used in spectroscopy, warfare, and astronomy.

These ranges are called atmospheric windows. There are specific atmospheric windows at infrared wavelengths between 730 nm and 100 mm. This wavelength range is called the near infrared, because it lies just beyond the red end of the visible spectrum.

Using infrared radiation

Applications using infrared radiation surround us in everyday life. Weather satellites use infrared imaging systems to map cloud patterns and make predictions about the weather. Similar systems are used to locate the edges of burning areas in forest fires, areas that are usually obscured by smoke. Infrared lamps are used to dry paint on production lines and printing ink on printing presses. Many restaurants use infrared lamps to keep food warm before it is served. Infrared lamps are also used as heaters in barns, sidewalk cafés, and other drafty areas where conventional heating would be impractical.

Infrared radiation has been important to astronomers (see ASTRONOMY). Infrared telescopes reveal information about the Universe that cannot be seen using ordinary visible light astronomy. The telescopes allow astronomers to view dust bands in our Solar System, dust disks around nearby stars, and distant galaxies that emit most of their radiation at infrared wavelengths. Making astronomical observations in the infrared spectrum can be difficult because the atmosphere tends to absorb infrared radiation. For this reason, infrared telescopes are usually located in the highest and driest places, such as on the mountains of Hawaii, where infrared absorption by atmospheric water vapor is minimized.

Infrared images cannot be captured on ordinary photographic film. The infrared radiation emitted by a body is recorded on a thermogram. Many thermograms are produced by infrared imaging systems that use light-sensitive crystals to generate electrical information that is then converted into an image. Others are made by false-color infrared photography, in which special films are designed to be sensitive to light in the near-infrared region of the spectrum. In false-color photographs, infrared, red, and green light appear as red, green, and blue, respectively.

Infrared spectroscopy is the study of how materials respond to exposure to infrared radiation. Every substance has a tendency to absorb and emit radiation at specific frequencies. These frequencies form a "fingerprint" by which an unknown material can be identified. Thus, infrared spectroscopy can be used to detect the presence of a substance and to tell how much of a given substance is present.

Infrared radiation also has applications in warfare, one of which is the Sidewinder heat-seeking missile. The latest, highly accurate missiles have detectors cooled to a very low temperature so that they can detect even the temperature difference between relatively cool areas of an aircraft and the surrounding air.

P. TESLER

See also: ELECTROMAGNETIC SPECTRUM; HEAT; SPECTROSCOPY; ULTRAVIOLET RADIATION.

Further reading:
Coletta, V. P. *College Physics.* St. Louis: Mosby Year Book, 1995.
Giancoli, D. C. *Physics: Principles with Applications.* Englewood Cliffs, New Jersey: Prentice-Hall, 1991.

This jet aircraft is carrying two Sidewinder heat-seeking missiles, which can detect the infrared radiation of enemy aircraft.

SEEING THE INVISIBLE

Infrared radiation is invisible. But certain devices help us "see" it. These infrared imaging devices convert an invisible infrared image into a visible image. The amount of infrared radiation that an object emits depends on the temperature of the object, as well as on the type of material it is made from. Infrared imaging devices "see" differences in amounts of infrared radiation emitted by different parts of an object and convert this information into a visible picture of heat distribution. The picture created can be enhanced electronically to convert certain temperatures into certain colors or to show only areas with a certain temperature range. Originally developed for military applications such as gun sights, infrared imaging is now also used for geological surveys, pollution monitoring, and finding points of excessive heat loss in buildings.

SCIENCE AND SOCIETY

INLAND SEAS

Inland seas are large enclosed bodies of water, often with high salinity

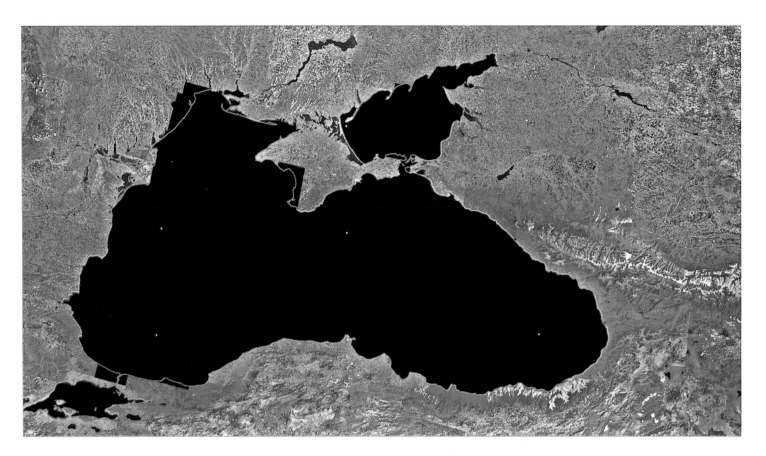

The Black Sea, located to the east of the Mediterranean Sea, is bordered by many nations: Ukraine, Russia, Georgia, Turkey, Bulgaria, and Romania. In the picture above, the Sea of Azov can be seen just above the center of the Black Sea, and the Sea of Marmara is to the bottom left.

CONNECTIONS

● The **SALT** content and size of a **WATER** body are often used as the determining factors between a **LAKE** and an inland sea.

● **CLIMATE** changes and resultant **DROUGHT** have led to the decline in size of many inland seas.

Although hydrologists may disagree on the exact definition of the term *inland sea,* it is generally agreed that it includes bodies of water that are surrounded by land except for a narrow strait that connects them to an ocean or another sea. They are found either on the continental shelf or on the interior of the continent. However, there are other large bodies of water that are generally classified as inland seas despite their lack of any link to another sea. While these may appear to be lakes (see LAKES), their size and the salt content of their waters place them in the inland seas category. The largest inland sea in the world, the Mediterranean Sea (see MEDITERRANEAN SEA), has an area of 1,145,000 sq miles (2,965,500 km²). Covered below are the other major modern inland seas, including the Caspian, Aral, Black, and Dead Seas.

Caspian Sea

The Caspian Sea is the second largest inland sea in the world today. In fact, at 152,239 sq miles (394,300 km²) in area, it is about five times the area of Lake Superior—the largest lake on the planet.

Located in southwestern Russia and northern Iran, the 745-mile (1200-km) length of the Caspian Sea presents a study in contrasts. For example, the southern areas are bordered by luxuriant vegetation that in the summer seems almost tropical. Meanwhile, the northern waters of the Caspian often remain frozen from December through March. Even in the middle of summer, this northern area is nearly without vegetation, and agriculture is difficult because of the salty soil.

The northern part of the sea is quite shallow, but central and southern regions have great depths, up to 3104 ft (946 m) at the deepest recorded site. The mean depth of the Caspian Sea is 597 ft (179 m). Salinity levels vary greatly throughout the sea and are strongly influenced by the prevailing counterclockwise current patterns. Despite the high level of biological activity in its upper levels, there is no life below the depth of 1500 ft (450 m) because of high concentrations of hydrogen sulfide (H_2S).

The Caspian lies 93 ft (28 m) below sea level, and its surface is continuing to decline. Despite the fact

CORE FACTS

■ Although disagreement still exists among scientists between the definitions for inland seas and lakes, inland seas are usually referred to as large enclosed bodies of water with high salinity.

■ The Mediterranean Sea is the largest inland sea.

■ The Dead Sea on the border of Israel and Jordan has such a high level of salinity that a human can easily float on its surface.

■ The high salinity of most inland seas results from high levels of evaporation and relatively low freshwater inflows.

that several rivers, including the Volga, flow into the sea, its high level of evaporation means that the sea could be 100 ft (30 m) below sea level within the next 40 years. Since the early 1960s, Russian engineers have worked on projects designed to divert water from other rivers into the Volga to increase the flow into the Caspian in an attempt to halt the steady decline in its water level.

In terms of human activity, the Caspian Sea closely resembles the open seas and oceans of the world. There is heavy shipping traffic throughout the sea, especially large tankers that carry oil from the refineries at Baku to the city of Astrakhan on the Volga. Baku itself, located on the shore of the Caspian, is one of Russia's largest and most important cities. Since the early years of the 20th century, large amounts of oil have been removed from wells drilled from trestles built out into the Caspian in the Baku region. Commercial fishing is also an important industry. One of the most important products of the Caspian is the world-famous caviar, which is obtained from sturgeon.

Aral Sea

The Aral Sea, which lies between Kazakhstan and Uzbekistan, is a good example of what can happen to an inland sea without proper ecological management.

In the early 1960s, the sea lay about 170 ft (51 m) above sea level. The presence in its waters of fish similar to those in the Caspian—just 175 miles (280 km) to the west—along with geological evidence indicate that the two seas were once linked. The salinity of the Aral's water was slightly lower than

> # TETHYS SEA
>
> The Mediterranean, Caspian, Aral, and Dead Seas were all once part of a single body of water—the Tethys Sea. This great ocean existed during the Mesozoic era (144 to 65 million years ago) and separated the ancient supercontinents Laurasia and Gondwana. (Laurasia was made up of North America and the northern part of Eurasia; Gondwana consisted of South America, Africa, India, Australia, Antarctica, and the southern section of Eurasia—see GONDWANA.)
>
> The existence of some sort of Mesozoic marine body where the European Alps and Himalayan mountain chain now stand was debated in the late 19th century. German geologist Melchior Neumayr found marine rocks and fossils in the mountains of the European Alps and Himalayas. In 1863, Austrian geologist Eduard Suess (1831–1914) stated that this evidence pointed to a former ocean, which he named Tethys. In the early part of the 20th century, however, many geologists disputed Suess's idea. They decided that the marine realm had in fact been a geosyncline—a narrow trough in Earth's crust. This view prevailed until the theory of plate tectonics became accepted (see PLATE TECTONICS).
>
> Scientists now believe there were at least two Tethys Seas during the Mesozoic era: the Paleo ("Old") Tethys Sea and the Neo ("New") Tethys Sea. The remains of the NeoTethys Sea, often called simply the Tethys Sea, form the eastern part of the Mediterranean Sea.

HISTORY OF SCIENCE

that of the Caspian and a thin layer of fresh water covered the entire 25,659 sq miles (66,457 km^2) of its surface. The Aral Sea was relatively shallow, with an average depth of 51 ft (15 m) and a maximum depth of 223 ft (67 m).

Only two rivers drain into the Aral—the Amu Darya and Syr Darya—and formerly the flow of water that they brought in was generally balanced

The drought in the Aral Sea has been so extensive that fleets of fishing boats have been left beached as the water level fell.

by loss from evaporation. There was some variation: over the centuries there had been changes in sea level of some 20 ft (6 m), and even seasonal variations could be as much as 10 ft (3 m). From 1960 onward, however, the Soviet government decided to divert large quantities of water from the two rivers for irrigation purposes, and the Aral Sea began to shrink rapidly. By the 1980s, it had lost more than half its water and was little more than half its previous depth. The rising salinity killed most of the fish, and the ports of Aral'sk and Muynak found themselves more than 30 miles (50 km) from the water.

By 1989, the level of the Aral had dropped about 50 ft (15 m), with the result that it became divided into two separate parts by the former seabed, which was now dry land. In 1992, the total area was calculated to be only 13,000 sq miles (33,800 km^2), and the salinity was three times what it had been 30 years before. In 1994, the states that relied on the water of the two rivers at last realized that something had

to be done, and they agreed to make efforts to increase the flow into the Aral. The most recent estimate of the total area covered by the Aral Sea is 14,024 sq miles (36,462 km^2).

Black Sea

Also located primarily in southwestern Russia, the Black Sea was connected to the Caspian Sea millions of years ago and together they formed a single sea. Today, the Black Sea has a surface area of 179,200 sq miles (423,000 km^2) and the only outflow is through the narrow Bosporus Strait into the Sea of Marmara and eventually into the Mediterranean Sea (see BLACK SEA). The Black Sea has an average depth of about 4000 ft (1200 m) and a maximum depth of 7250 ft (2200 m). The depth, as well as the salinity, of the Black Sea fluctuated widely throughout its geologic history but has remained stable for the past 5000 years.

The large freshwater inflow from numerous rivers, including the Danube, results in a layering effect

Crystalline salt formations in the Dead Sea result from the inflow of dissolved salts, extremely high rates of evaporation, and no outflow.

between fresh water and the heavier salt water in the Black Sea. Considerable mixing of the two occurs at the Bosporus, where salt water flows in at depth and fresh water passes out at the surface. Despite the fact that differences in densities keep the fresher water at the surface and the saltier water at greater depths, a large amount of salt water comes to the surface each year. Oxygen supports life down to a depth of 660 ft (200 m), but hydrogen sulfide appears in heavy concentrations beyond this depth and only certain bacteria can survive in this zone.

Although considered by many hydrologists to be a gulf or lagoon of the Black Sea, the Sea of Azov also can be classified as an inland sea. It is connected to the Black Sea by the narrow and shallow Kerch Strait and has a mean depth of only 16 ft (5 m). The Sea of Azov is 14,517 sq miles (37,750 km^2) and has a maximum depth of only 45 ft (13.5 m). Only about 100 different species of plants and animals inhabit the Sea of Azov. In contrast, the Black Sea has some 1200 species—while about 7000 species live in the Mediterranean.

Dead Sea

The Dead Sea is located on the border between Jordan and Israel and covers 538 sq miles (1393 km^2). During the Jurassic and Cretaceous periods, the Dead Sea was part of the Mediterranean Sea, which at that time covered Syria and Palestine.

Although a number of small streams flow into it, the Dead Sea has no outlet. In addition to its role in Biblical history, the Dead Sea is also notable because of its extreme salinity (as much as seven times saltier than ocean water) and its location, 1302 ft (396 m) below sea level—the lowest body of water in the world. The Dead Sea is deepest, at 1200 ft (360 m), in its northern region, but becomes very shallow at its southern end. The average depth of its waters is 460 ft (138 m).

The salinity increases with depth. The upper 130 ft (40 m) of water has a salt concentration of just under 300 parts per thousand. The lower waters are saltier, with a salinity of 332 parts per thousand, and contain high levels of magnesium, potassium, chlorine, and bromine. Because of its density, the lowest water remains permanently at the bottom.

The name *Dead Sea* was derived from the belief that the sea was too salty to support life. While this saltiness does preclude the existence of the fish and other types of life associated with seas, several lifeforms (mainly bacteria) have been found within the Dead Sea's waters, and large numbers of birds, animals, and reptiles live along its shore. The salt content of the water makes the water very dense and creates sufficient buoyancy to keep a human body on the surface. Swimmers float easily.

The Dead Sea plays a prominent role in both the Old and New Testaments of the Bible. Sodom and Gomorrah, the cities whose evil inhabitants were reportedly killed by God, have been placed in the vicinity of the Dead Sea.

J. HUMPHREYS

GREAT SALT LAKE

Located in Utah in the western United States, Great Salt Lake has many of the same characteristics of the Dead Sea. However, it is classifed as a lake because it was never linked to a sea or ocean, even in prehistoric times, and many hydrologists classify it as an inland lake. The water is, as the name implies, extremely salty. In fact, the waters of Great Salt Lake are second only to the Dead Sea in terms of salt content. The lake is actually the dying remains of Lake Bonneville, a huge prehistoric lake, which was as much as 1000 ft (300 m) deep. As this lake slowly evaporated, it left behind high concentrations of salt and other minerals.

Today, Great Salt Lake averages 13 ft (4 m) in depth and has a maximum depth of 22 ft (6.7 m). Determining the actual surface area is a difficult task, because it can change dramatically from year to year, based on the rate of inflow from three rivers and the rate of evaporation. For example, in 1873, the lake covered 2300 sq miles (5980 km^2). By 1940, the lake was 1078 sq miles (2800 km^2)—its smallest recorded area. Since then, it has experienced a gradual increase to about 1800 sq miles (4660 km^2). Despite its salinity, Great Salt Lake provides a home for brine shrimp, and many of the islands in the lake are inhabited by pelicans and gulls.

Great Salt Lake as seen from space. The lake can be seen in the foreground. Beyond the lake is Great Salt Lake Desert.

A CLOSER LOOK

See also: BLACK SEA; LAKES; MEDITERRANEAN SEA; RIVERS.

Further reading:
Corso, W. *Oceanography*. Springhouse, Pennsylvania: Springhouse Corporation, 1995.
Earle, S. A. *Sea Change: A Message of the Oceans*. New York: G. P. Putnam's Sons, 1995.
Richie, D. *The Inland Sea*. New York: Kodansha International, 1993.

IONOSPHERE

The ionosphere is the upper layer of Earth's atmosphere and contains free electrons and ions

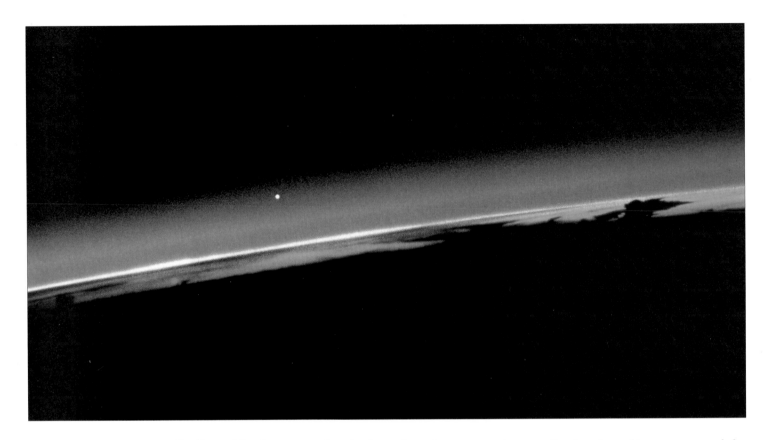

This view of Earth's atmosphere, seen at sunset, was taken from the space shuttle Atlantis. *The blue is caused by reflections of the Sun's light in the atmosphere. The orange glow is from clouds seen just below the horizon.*

CONNECTIONS

● The **ENERGY** required to remove an **ELECTRON** from a free **ATOM** or **ION** in the gaseous state is called the ionization energy.

● During a solar **ECLIPSE**, the intensity of ionizing radiation drops abruptly.

Most of the shorter wavelength ultraviolet radiation from the Sun is filtered out by the highest layers of Earth's atmosphere. Solar radiation, streaming from the Sun in the solar wind, ionizes some of the atmospheric molecules, creating a constantly changing mix of molecules with ions and electrons. This is the ionosphere, which extends from about 30 miles (50 km) above Earth's surface to as much as 620 miles (1000 km).

The ionosphere is dynamic. Not only does it vary with day and night and the season of the year, it also responds to the sudden darkening of solar eclipses (see ECLIPSES). It changes with differences in the solar wind caused by solar flares (see SUN). Ionospheric research was very important when long-distance communication and broadcasting was done by shortwave radio. Now, satellites (see SATELLITES) and optical fiber undersea cables (see FIBER OPTICS) carry most of this traffic.

A new age in communications

The ionosphere is essential to long-distance radio communication. Radio waves (see RADIO AND RADAR), like light waves, travel in straight lines (this is known as line of sight); so, just as we cannot see what is beyond the horizon, we would expect radio waves to shoot straight out into space.

In 1899, however, Italian scientist Guglielmo Marconi (1874–1937) showed that radio transmissions could be sent more than 60 miles (100 km) over the ground. This is because the ionosphere acts as a mirror, reflecting the radio waves around the curve of Earth (see REFLECTION AND REFRACTION).

Because the ionosphere is ionized, it is a good electrical conductor. As early as 1839, scientists began to speculate that there was an upper layer of the atmosphere that would conduct electricity.

In 1902, following Marconi's success with long-range radio, U.S. engineer Arthur Kennelly (1861–1939) announced the probable existence of a region of ionized air favorable to radio-wave propagation. In the same year, British physicist Oliver Heaviside (1850–1925) also showed how the existence of an ionized atmospheric layer would explain why radio waves were transmitted around Earth. Proof of the existence of the then-called Kennelly-Heaviside layer was made in 1924 by English physi-

---

CORE FACTS

■ The ionosphere, the upper layer of Earth's atmosphere, contains free ions and electrons produced by the Sun's ultraviolet radiation and X rays.

■ There are four layers in the ionosphere: the D layer begins at a height of 30 miles (50 km); the E layer at 55 miles (90 km); the F_1 layer at 90 miles (140 km); and the F_2 layer at 125 miles (200 km).

■ Without reflection by the ionosphere, long-distance radio transmission at low frequencies would be impossible.

■ Because ionization is produced by solar radiation, some of the layers weaken or disappear during the night.

cist E. V. Appleton (1892–1965) who, among others, compared the fading of signals in two different directions. Appleton also established the existence of another layer outside the Kennelly-Heaviside layer, called the Appleton, or F_2, layer. Appleton showed that radio waves of sufficiently short wavelength would penetrate the lower region of the ionosphere and be reflected by an upper region, now called the F_2 layer. This discovery led to improvements in the reliability of long-range radio communication.

The structure of the ionosphere

Although the ionosphere is constantly changing, it nonetheless has a layered structure of distinct regions. These layers develop because both energy input from the Sun and the composition—and hence the density—of the atmosphere vary with altitude. These atmospheric differences affect the rate at which ions can recombine to form neutral molecules. There are four ionospheric layers that can be roughly described by altitude (or temperature), but they are better described by the physicochemical processes that take place in each.

Four distinct regions

The lowest region, called the D region, occurs at 30 to 55 miles (50 to 90 km). Here, extreme ultraviolet rays and weak X rays from the Sun produce ionization of nitric oxide (NO), negative ions being formed by the combination of neutral molecules and electrons. Negative ions are also produced in the lower D layer due to ionization by cosmic radiation (see COSMIC RADIATION).

In the E region, at 55 to 90 miles (90 to 140 km), ionization of oxygen molecules (O_2) is the primary process. This, too, is driven by extreme ultraviolet and weak X rays but may also be caused by meteors (see METEORS AND METEORITES). In the F_1 layer, at 90 to 125 miles (140 to 200 km), oxygen molecules (O_2) are ionized.

Since the ionization processes are driven largely by solar radiation, the D layer disappears and the F_1 layer merges with the F_2 soon after sunset, and the E layer weakens considerably. The F_2 layer, at 125 to 300 miles (200 to 480 km), persists but becomes smaller, reaching a minimum just before sunrise.

Above the F_2 layer is the magnetosphere, the region of near-Earth space in which Earth's magnetic field controls the behavior of charged ions and electrons. The magnetopause is the boundary surrounding the magnetosphere. Here the force of Earth's magnetic field is balanced by the force of the solar wind. Normally, solar wind plasma flows around the magnetosphere and along the magnetopause, leaking only small amounts of radiation into Earth's atmosphere. But during a magnetic storm (a disturbance in Earth's magnetic field), large amounts of solar radiation cascade into the upper atmosphere.

In the ionosphere, very few atmospheric molecules are actually ionized—barely 0.4 percent of the population present at any altitude. But this seemingly slight "seasoning" of ions and electrons has

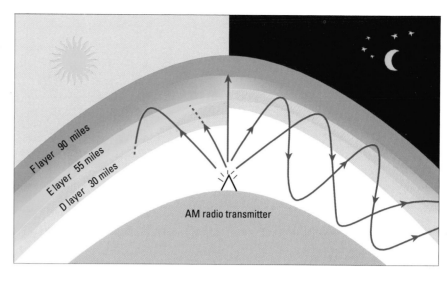

some very significant effects. Ionospheric changes will cause fluctuations in the magnetism at any point on Earth's surface.

Radio signals

The radio frequency reflected by an ionospheric layer is related to the density of free electrons, so sudden changes in a layer can result in unexpected strengthening or loss of radio signals. Radio stations constantly compensate and adjust their outgoing signals to follow the changing electrical characteristics. The effect is only dramatic in the lower (D and E) layers. At F_1 and above, the layers are relatively consistent, and variations are quite subtle. This effect is particularly important at sunset and sunrise.

Just how much a radio wave is reflected by the ionosphere depends on the frequency of the radio transmission. Low-frequency (longwave) transmissions, such as AM radio, are much more affected than high-frequency (shortwave) transmission such as FM radio or TV. Because of this, AM stations may reduce power or go off the air at night to avoid interfering with other stations hundreds of miles away.

Long-range radar, which returns a signal to the sender to indicate location of a target, can be similarly affected. Distortion of the signal path can indicate a totally erroneous position of the target. The ionosphere can cause scintillation—a rapid, usually random variation in signal caused by abrupt variations in electron density along the signal path, similar to the twinkling of a star.

P. WEIS-TAYLOR

See also: ATMOSPHERE; ELECTRONS AND POSITRONS; IONS AND RADICALS; RADIO AND RADAR; REFLECTION AND REFRACTION; STRATOSPHERE; SUN.

Further reading:
Davies, K. *Ionospheric Radio*. Waltham, Massachusetts: Blaisdell, 1993.
National Research Council. *The National Geomagnetic Initiative*. Washington, D. C.: National Academy Press, 1993.
Yeh, K. C. and Liu, C. H. *Theory of Ionospheric Waves*. New York: Academic Press, 1972.

At night, higher regions of the atmosphere transmit AM radio waves for many hundreds of miles. During the day, lower regions of the atmosphere strongly absorb and weaken the AM radio waves, preventing them from being picked up by distant receivers.

IONS AND RADICALS

Ions are atoms or molecules with missing or extra electrons; radicals are those with one unpaired electron

Electrons are gained and lost through a number of different processes. They can be transferred from one atom to another, such as when sodium and chlorine react to form sodium chloride (see CHEMICAL REACTIONS), the ionic compound we know as table salt. In other reactions, ions are created by proton transfer. For example, when a molecule of acetic acid (CH_3COOH) comes in contact with water, it may transfer a proton (a hydrogen nucleus) to the water in the following reaction, where CH_3COO^- are acetate ions and H_3O^+ are hydronium ions:

$$CH_3COOH + H_2O \rightarrow CH_3COO^- + H_3O^+$$

Ions can also be created by the bombardment of atoms or molecules with high-energy electrons or protons.

Electrolysis
In the solid state, ions exist as components of ionic crystals. These are orderly lattice structures of ions held together by the mutual attraction of the positive and negative ions in the lattice. Because ions occupy fixed locations in the crystalline lattice, they cannot conduct electricity easily.

When the ionic crystal is melted or dissolved in solution, the ions become mobile and so are able to conduct electricity. A substance that turns into ions, partially or completely, when dissolved in solution or melted acts as a conductor of electricity and is called an electrolyte. Electrolytes used in solution dissolve only in polar solvents—that is, those that contain molecules with positive and negative ends—such as water (H_2O). The water molecule is electrically neutral, carrying no electric charge. However, its oxygen side has a slight negative charge, while the two hydrogens have a slight positive charge. This polar concentration of charge acts to pull solid electrolytes apart. Once in solution, the water (or other solvent) molecules tend to stick to the ions of the electrolyte.

The reactions that produce this smog over Los Angeles involve radicals.

CONNECTIONS

● Radicals are quick to undergo **CHEMICAL REACTIONS** with single atoms.

● During **ELECTROLYSIS**, ions in a solution migrate to electrodes.

We come across things that contain ions every day. Common table salt contains sodium and chloride ions. Potassium ions in the nerves in our body allow them to conduct electrical impulses. Hydrogen ions are contained in the acids we use in the laboratory.

Ions result when atoms or groups of atoms lose or gain electrons. An ion with missing electrons has a positive charge and is called a cation. One with extra electrons has a negative charge and is called an anion. Most ions formed from a hydrogen atom or a metal atom are cations, while most formed from an atom of a nonmetal or a group of nonmetal atoms are anions. The number of missing or extra charges carried by an ion is called its electrovalence. This is written as a superscript that gives the type and number of charges. For example, a calcium ion has two missing electrons and is written as Ca^{2+}.

CORE FACTS
■ The loss or gain of electrons by an atom or molecule creates charged particles called ions.
■ Ionic compounds contain mobile electrons when molten or when dissolved in a polar solution; these substances are called electrolytes.
■ Electrolysis is the chemical process by which an electrolyte is partially or fully decomposed by an electrical potential difference.
■ The solubility product is a limit for the value of the product of ion concentrations in solution; beyond the limit, a solid precipitate is formed.
■ Radicals are highly reactive, short-lived atoms or molecules with one unpaired electron.

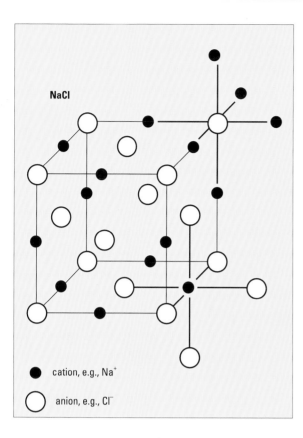

NaCl

● cation, e.g., Na$^+$

○ anion, e.g., Cl$^-$

The ions in solid sodium chloride, as in other ionic compounds, are held together in an orderly lattice structure.

The process of decomposing an electrolyte using a potential difference is called electrolysis. In electrolysis, a battery is connected to two conductive electrodes, which are submerged in the electrolyte. The battery creates a potential difference across the electrodes, causing oppositely charged ions in the solution to move in opposite directions. Positive ions migrate to the negative electrode (cathode), while negative ions migrate to the positive electrode (anode). At the electrodes, chemical reactions occur to complete the circuit. During these reactions, electrons are either accepted (reduction) or donated (oxidation). These reactions are used in the production of chemicals, extraction of metals, electroplating, and for the production of electricity in batteries.

Acids and bases are two important kinds of solutions that can be described in terms of ions. In general, acids raise the concentration of H$^+$ ions in a solution and bases raise the concentration of OH$^-$ ions in solution. A base added to an acidic solution is said to neutralize the solution, so that neither positive nor negative ions predominate.

Salt solubility

Some ionic solids have only limited solubility. The solubility product is a numerical constant that specifies the point at which a solution can no longer hold additional ions. Such solutions are said to be saturated. For example, when solid silver chloride (AgCl) is added to water, it dissociates to silver ions (Ag$^+$) and chloride ions (Cl$^-$) in the following reaction:

$$AgCl \rightleftharpoons Ag^+ + Cl^-$$

Silver chloride continues to dissociate until the following condition is met, where [Ag$^+$] is the concentration of silver ions, [Cl$^-$] is the concentration of chloride ions, and K is the constant called the solubility product:

$$[Ag^+] \times [Cl^-] = K$$

Silver chloride will dissolve until the concentration of silver ions multiplied by the concentration of chlorine ions equals the solubility product. There is no specific restriction on the concentrations of each ion, but the product of their concentrations cannot exceed the solubility product. If the product of ion concentrations is greater than the solubility product, then the ions do not dissolve in the saturated solution and a solid precipitate is produced. The solubility product is unique for each ionic compound. It also varies with the solvent and the temperature.

The solubility product is determined experimentally for each ionic solid dissolved in water. However, if ionic solids are dissolved in solvents containing a common ion, they may prove to be less soluble. One of the causes for this is known as the common-ion effect: the solubility of an ionic solid decreases in

When acid from the burette has completely neutralized the alkali in the flask, the indicator will turn from purple to colorless. At this point, the number of hydrogen ions in the solution will equal the number of hydroxyl ions.

Complex ions

Some ions, called complex ions, consist of a central metal atom surrounded by a tightly bound first layer of atoms and a loosely bound second layer. The inner layer is called the first coordination sphere and the outer layer is called the second coordination sphere. Atoms or molecules that stick to the central metal atom are called ligands (from the Latin word *ligare*, meaning "to tie"). The number of ligands that can be accommodated in the first coordination shell is called the coordination number of the complex ion. Compounds that contain complex ions are often called coordination compounds.

One example of a complex ion is $Cr(NH_3)_6^{2+}$, (see the diagram at left), also known as hexamine chromium. This complex ion is composed of a central chromium (Cr) atom surrounded by six molecules of ammonia (NH_3). Thus, the coordination number of the ion is six.

Complex ions play important roles in many chemical and biological systems. They are the structural units in many crystals and are often involved as intermediates in the mechanisms of chemical reactions. Naturally occurring complexes include hemoglobin, the iron-containing protein in mammalian blood that binds with oxygen, and chlorophyll, the magnesium complex that is involved in photosynthesis in plants.

the presence of a solvent that has an ion in common. For example, silver sulfate (Ag_2SO_4) dissolves more readily in pure water than in a sodium sulfate (Na_2SO_4) solution. This is because when silver sulfate is mixed with sodium sulfate, both substances dissociate to form the sulfate ion (SO_4^{2-}), so only a very small Ag^+ concentration is needed for the product $[Ag^+]^2[SO_4^{2-}]$ to exceed the solubility product. The sulfate ion concentration in a sodium sulfate solution is already substantial. When silver sulfate is added, it produces more sulfate ions. Hence, less of it can dissolve without saturating the solution.

This hexamine chromium ion is made up of a central chromium atom surrounded by six molecules of ammonia.

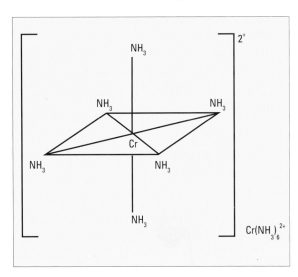

Radicals

An atom or molecule that has one unpaired electron is called a radical. Usually, molecules contain an even number of electrons, and the covalent bonds that hold them together consist of shared pairs of electrons. Radicals are often the result of a cleavage of normal electron-pair bonds, resulting in two atoms or molecules, each with one unpaired electron. These species are called radicals, and they are usually electrically neutral. When they are neutral, the total number of electrons in the radical can just balance the total positive charge of the nuclei. Radicals can also be ionized.

As a result of having an unpaired electron, radicals tend to be highly reactive. They are generally quick to combine with each other or with single atoms, forming molecules with the normal paired electron arrangement. Radicals can also react with molecules that do not contain unpaired electrons. In these reactions, the radical "robs" the molecule of parts that will complete its own electron pairs, leaving behind another radical in its place.

Because they are often so highly reactive, most radicals do not exist independently for very long. In fact, radicals have such a short life that for many years scientists believed that they did not, and could not, exist. This belief was changed in 1900 by U.S. chemist Moses Gomberg (1866–1947). While he was trying to synthesize the organic chemical hexaphenylethane, Gomberg discovered the triphenylmethyl radical ($C_6H_5)_3C$. Gomberg's discovery was not immediately accepted by the scientific community. However, many other radicals have been discovered since, and their existence is no longer debated.

Although many radicals are highly reactive, certain ones are stable and can exist independently for a long time. Triphenylmethyl is one example, as is nitric oxide (NO).

Radicals can be created in a number of ways. Some molecules that are weakly bonded can be broken down into radicals by heating, a process called thermal decomposition. Electrical and microwave discharge, as well as electrolysis, can also generate radicals. They can also be created by irradiation with various forms of light. In these reactions, called photochemical reactions, the light imparts sufficient energy to break chemical bonds. Most photochemical reactions involve ultraviolet light (see ULTRAVIOLET RADIATION), but some molecules, such as nitrogen dioxide (NO_2, a brown gas found in smog) and molecular chlorine (Cl_2, used in swimming pools) will decompose into radicals in visible light. Almost all photochemical reactions are thought to involve radicals at some stage. Radicals are also known to play a role in high-temperature reactions.

Chain reaction

Radical reactions are often chain reactions, that is, reactions in which one of the products is also one of the reactants. An example is the reaction of methane (CH_4) and chlorine (Cl_2) to yield methyl chloride (CH_3Cl) and hydrogen chloride (HCl). The reaction involves the following steps:

1. $Cl_2 \rightarrow 2Cl$
2. $Cl + CH_4 \rightarrow CH_3 + HCl$
3. $CH_3 + Cl_2 \rightarrow CH_3Cl + Cl$
4. $2Cl \rightarrow Cl_2$

Step 1 is the initiation, steps 2 and 3 are called propagation, and step 4 is the termination reaction. The two products of the reaction, hydrogen chloride (HCl) and methyl chloride (CH_3Cl), are produced in the second and third steps, and the chlorine molecules (Cl_2) that initiate the reaction in the first step are regenerated in the fourth. In this way, a single atom of chlorine can participate in the production of an endless number of methyl chloride and hydrogen chloride molecules. Chain reactions like this one are involved in making plastics (see PLASTICS), as well as in the reactions that take place in the atmosphere, where short wavelength ultraviolet rays cause oxygen molecules to break apart and form oxygen radicals. The oxygen radicals react with oxygen molecules to form ozone (see OZONE).

Radicals have unique magnetic properties as a result of having unpaired electrons. Molecules with paired electrons are repelled by a magnet and are known as diamagnetic materials (see MAGNETISM). Radicals are attracted by a magnet and are called paramagnetic materials. Paramagnetism can be measured in terms of the force on a sample in an external magnetic field, also known as the magnetic susceptibility.

P. TESLER

See also: ACIDS AND BASES; CHEMICAL REACTIONS; ELECTROLYSIS; SALTS; TRANSITION ELEMENTS.

ION EXCHANGE

Hard water is water that contains an above normal amount of calcium and magnesium salts. It forms a deposit of carbonate salts, known as scale, inside kettles, boilers, and pipes and a scum with soap. Water softening is a chemical process called ion exchange. This is the process of exchanging ions from a solid phase with ions in solution. A water softener removes the calcium and magnesium ions from hard water and exchanges them for other ions, such as sodium ions.

Many natural and human-made substances are used in ion exchangers, including silicate minerals, synthetic resins, proteins, bone, and even living cells. The surfaces of these solids contain electrically charged sites. These sites pair up with simple ions of opposite charge, and it is these ions that are exchanged with the ions from solution.

Ion exchange is used not only for water softening, but also for chemical analysis and ion separation for industrial purposes. For example, in sugar refining, ion exchangers remove the naturally occurring salts that keep the sugar from crystallizing rapidly and completely.

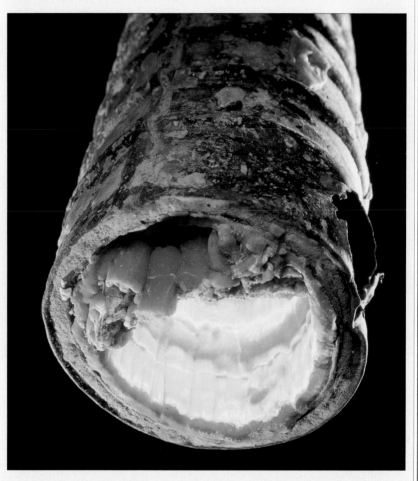

Thick deposits of lime scale inside domestic metal pipes are avoided by using a water softener, which has an ion exchange column to filter out the ions that form such deposits.

A CLOSER LOOK

Further reading:

Barrett, J. R. *Understanding Inorganic Chemistry*. New York: Ellis Horwood, 1991.
Dickson, T. R. *Introduction to Chemistry*. New York: John Wiley & Sons, 1995.
Hazen, R. M. and Trefil, J. *Science Matters*. New York: Doubleday, 1991.

IRON AND STEEL

Iron is a gray, shiny transition metal that is used to manufacture steel

CONNECTIONS

● Tin cans are made by coating iron with a layer of **TIN**.

Symbol: Fe
Atomic number: 26
Atomic mass: 55.847
Isotopes:
  54 (5.82 percent),
  56 (91.2 percent),
  57 (2.10 percent),
  58 (0.28 percent)
Electronic shell
  structure: [Ar]3d^{6}4s^2

Fragments from meteorites containing metallic iron can be found all over the world. Ancient peoples probably first became aware of iron from meteorites they found. It is possible that, when they discovered that this metal could be worked by heating it, they began to look for other similar substances and discovered the ore now known as hematite; this is hard and heavy and has a rusty-red metallic appearance. It is an oxide of iron, but when it is heated with charcoal it gives lumps of iron that can be hammered into shape.

The ancient Romans valued iron and traveled as far as Britain to find sources. The chemical symbol, Fe, comes from the Latin word *ferrum*. Iron's common name derives from the Anglo-Saxon word *iren*.

Properties of iron

Pure iron is a gray, shiny metal with a melting point of 2800°F (1536°C), but it is not particularly hard. However, it almost always contains small quantities

The oxidation of iron and certain other metals appears as a brown, flaky layer of rust.

of carbon, which greatly affect its properties, making it harder and stronger but lowering its melting point considerably.

As one of the elements of the first transition series (see TRANSITION METALS), iron is grouped with cobalt and nickel—forming a triad—in group 8 (VIII) of the periodic table. Iron has atomic number

CORE FACTS

▪ Earth's core is composed mainly of molten iron and nickel.
▪ Iron is ferromagnetic below 1410°F (766°C).
▪ Industrially, iron is manufactured by the reduction of its oxides in a blast furnace.
▪ Steel is any alloy of carbon and iron containing 0.05–1.6 percent carbon.

26, with electron configuration 2, 8, 14, 2. It forms compounds in two main oxidation states, +2 and +3. A higher oxidation state, +6, is also known, but it is not particularly stable.

Iron is more strongly affected by magnets than any other element. Once magnetized, it will retain its own powerful magnetism—it is said to be ferromagnetic. However, if it is heated above 1410°F (766°C), magnetized iron will lose its magnetism and become only paramagnetic—that is, it can be only weakly magnetized and only as long as it is in a magnetic field. Cobalt and nickel are the only other ferromagnetic elements at room temperature (see MAGNETISM).

Ferromagnetism occurs in solid iron, cobalt, and nickel because their individual paramagnetic atoms are spaced just the right distance apart; they are able to line up with one another and form strong magnetic regions called domains. Placed in a magnetic field, these domains then also line up together; and, when the field is removed, they stay aligned, making a permanent magnet. Ferromagnetism is about one million times as strong as paramagnetism.

A permanent magnet is not really permanent, however. The domains can be disordered by impact—hitting with a hammer, for example—or by heating them.

Compounds of iron

Compounds of the +2 oxidation state are called ferrous; those of the +3 oxidation state are called ferric. Three oxides are known: FeO, Fe_2O_3, and Fe_3O_4.

Ferrous oxide, FeO, is difficult to prepare. It is a black powder that burns, often spontaneously, in air. The product of combustion is red-brown ferric oxide—Fe_2O_3—the most common iron oxide and the constituent of rust (see the box on page 619). The black mineral magnetite is Fe_3O_4. In this compound, iron is present in both +3 and +2 states.

Iron forms compounds with many nonmetals and reacts with acids to give a variety of salts. Most iron salts, both ferrous and ferric, are colorless in the pure state, but iron easily forms complexes with water, so many crystals and solutions in water are colored: green in the case of ferrous salts, yellow in the case of ferric salts.

The best known ferrous salt is the sulfate, $FeSO_4$, which is sometimes called green vitriol. It occurs in nature as the mineral melanterite, or copperas, which is the result of oxidation of pyrite, ferrous sulfide (FeS_2). It was used by the Greeks and Romans to make ink: adding ground oak galls produced the intensely black iron gallate. Ferrous sulfide is used in the laboratory for the preparation of hydrogen sulfide gas (H_2S) by reaction with dilute acids.

Ferric salts in solution are yellow, but they turn a darker color due to the formation of brown ferric hydroxide, $Fe(OH)_3$. Ferric sulfate, $Fe_2(SO_4)_3$, forms alums with the sulfates of the alkali metals. Potassium iron alum, $K_2SO_4.Fe_2(SO_4)_3.24H_2O$, will form pale violet crystals of great purity. $Fe_2(SO_4)_3$ readily decomposes to Fe_2O_3 and SO_3 on heating. A sure test for the presence of Fe^{3+} ions is to add a

The difference in color between potassium ferrocyanide (left) and potassium ferricyanide (right) is due to the differing oxidation state of iron. In potassium ferrocyanide, iron is in the +2 oxidation state. In the ferricyanide, iron is in the +3 oxidation state.

solution of potassium thiocyanate (KSCN). If Fe^{3+} ions are present, a deep red color is produced due to the formation of the hydrated thiocyanate complex $[Fe(CNS)(H_2O)_5]^{2+}$. No coloration is produced with Fe^{2+} ions.

Somewhat similar to the alums are the ferrocyanides and ferricyanides. Potassium ferrocyanide, $K_4Fe(CN)_6$, forms large yellow crystals; potassium ferricyanide, $K_3Fe(CN)_6$, is deep red. When solutions of ferrocyanide and a ferric salt are mixed, a deep blue precipitate called Prussian blue is obtained; mixing ferricyanide and a ferrous salt produces Turnbull's blue. These two solids, which contain a mixture of ferrous and ferric compounds, are in fact identical. Prussian blue has been widely used as a pigment, and its formation when ferro- or ferricyanide is added to an unknown solution is a sure analytical test for iron and gives an indication of its oxidation state. Like the other transition metals, iron will react directly with carbon monoxide to form a carbonyl, $Fe(CO)_4$, in which it is in a zero oxidation state.

Occurrence of iron

Iron is the fourth most common element on Earth, making up about 5 percent of Earth's crust, and is the second most abundant metal after aluminum. Earth's core is thought to be composed mostly of molten iron and nickel (see EARTH, STRUCTURE OF).

The principal ores are the red-brown hematite (largely Fe_2O_3) and black magnetite (mainly Fe_3O_4). In the United States, the major sources of these ores have been overexploited, and attention has recently turned to taconite, a form of magnetite together with silicates. Other fairly rich ores are the ferrous carbonates, but these generally occur mixed with a large proportion of impurities such as clay. Pyrite is the disulfide (FeS_2). Known as fool's gold because of its goldlike appearance, it is an important source of both ferric oxide and sulfur dioxide. Another crystalline form of pyrite is marcasite; it is not as brassy in appearance and is often used for cheap jewelry. Because iron is so widespread, it is rare to find an ore or mineral that does not contain some traces of it.

Iron is also a component of living systems. Heme is an iron-containing part of many important proteins, including hemoglobin, the essential oxygen-carrier in animal blood; myoglobin, responsible for oxygen transport in muscles; and the cytochromes, which occur in almost all organisms (see PROTEINS).

Manufacture of iron

The primitive production of iron from its oxides did not involve melting the metal. The ore was mixed with charcoal and the pile set on fire. The carbon of the charcoal reacted with the oxide, carrying off carbon monoxide:

$$Fe_2O_3 + 3C \rightarrow 2Fe + 3CO$$
$$Fe_3O_4 + 4C \rightarrow 3Fe + 4CO$$

A lump of relatively pure iron, called a bloom, was left. The bloom could be hammered into shape while red-hot and was therefore called wrought iron (*wrought* means "to work").

The blast furnace is used to reduce iron oxides to molten iron.

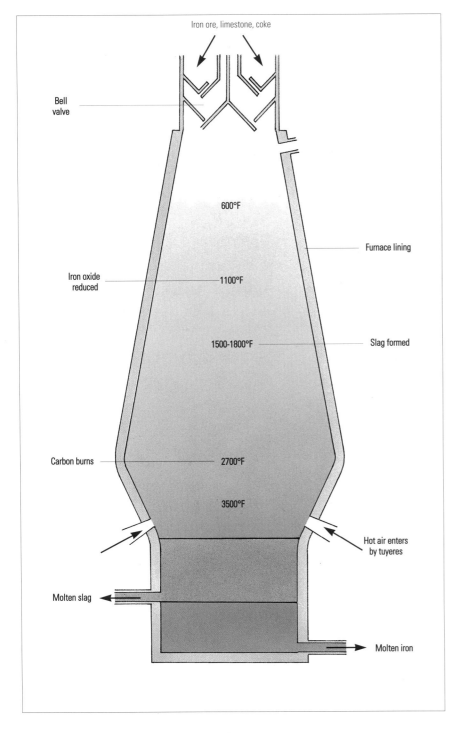

Iron ore, limestone, coke

Bell valve

600°F

Furnace lining

Iron oxide reduced — 1100°F

1500-1800°F — Slag formed

Carbon burns — 2700°F

3500°F

Hot air enters by tuyeres

Molten slag

Molten iron

WHY MAGNETIC?

Not only metallic iron is magnetic: the ore magnetite (Fe_3O_4) also has its atoms so spaced that it responds to a magnetic field. This mineral forms black crystals, or massive lumps, and because it could be hung from a thread to point north and south, it was known in the Middle Ages as lodestone, from the old word *lode* meaning "way" or "course." The mineral is called magnetite because it was discovered in an area in Asia Minor that the Greeks called Magnesia.

During the Middle Ages, special blast furnaces were developed that were hot enough to melt the iron, which could then be poured into molds; this was called cast iron. In the mold it formed a central bar, with side bars attached along both sides. It looked like a mother pig nursing her litter, so the cast bars were called pigs.

Modern blast furnaces use the same principle: the reduction of iron oxide to iron by heating the oxide over carbon. Iron ore, coke, and limestone are added to the top of the furnace. This mixture is called the charge. The coke reduces the iron oxide to the metal. Impurities in the ore react with the limestone to form slag. This has a lower melting point than the impurities and can be drained off at the bottom of the furnace. Molten iron trickles down the furnace and collects at the base.

Cast iron was much harder than wrought iron, but it was brittle because it contained a lot of carbon, so it could not be hammered. Cast iron could be changed into wrought iron by burning away excess carbon on a hearth, but this was a slow process. In 1784, Englishman Henry Cort (1740–1800) patented his puddling furnace. This was a reverberatory furnace. The fuel was burned at the rear, and the flames passed under a low roof to melt pig iron at the front, so the iron did not come into contact with the fuel.

As the carbon content of the iron was burned out, the melting point rose, so the iron became semi solid. Through a small side door to the furnace, a worker (the puddler) had to keep the mass stirred and form it into blooms, each weighing about 100 lb (46 kg). This iron was almost pure and free of carbon. It was used to make steel.

Manufacture of steel

The iron produced in a blast furnace contains around 3 to 4 percent carbon, which makes it too brittle for most uses. But a smaller proportion of carbon makes iron harder and stronger. This iron-carbon alloy is called steel.

Since 1953, steel has been produced using the basic oxygen process, which uses a furnace called the basic oxygen converter (see the diagram at right). The converter is held on a pivot, so it can be tilted. At the start of the process, the converter is tilted and molten iron is poured in. Oxygen and powdered lime are blown onto the surface of the iron through a water-cooled lance. This causes impurities in the iron to become oxidized, and the oxides react with the lime to form slag. The converter is tilted again to remove the slag, and the molten steel is collected.

The basic oxygen process makes carbon steel, which has a carbon content of up to 1.5 percent. This is the simplest type of steel and the cheapest to produce. It is used to make automobile bodies, ships, and machinery.

Alloy steels contain small amounts of other metals, such as nickel, chromium, cobalt, aluminum, and titanium. The different metals give the alloy steel a particular function. Nickel, for example, makes the

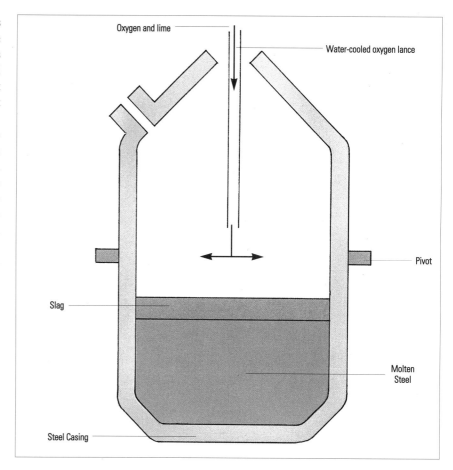

steel resistant to pulling stresses; tungsten steel is for tools that are used at very high temperatures.

E. KELLY

The basic oxygen converter is used to manufacture high-quality steel.

See also: ALLOYS; MAGNETISM; METALS; NICKEL AND COBALT; SMELTING; TRANSITION ELEMENTS.

Further reading:
Kirshner, R. P. "The Earth's Elements." *Scientific American*, **271**, p.58, Oct. 1994.

RUST

Why does the brown flaky material that spoils the shine of bicycles appear in the first place? Rusting is a form of corrosion. Rust develops when iron and some other metals come into contact with oxygen, acid (for example, in acid rain), or other kinds of substances found in the environment and sparks an electrochemical reaction.

Iron is more likely to be oxidized (that is, lose electrons; see OXIDATION-REDUCTION REACTIONS) than hydrogen or oxygen. It tends to act as a reducing agent, becoming oxidized itself:

$$Fe \rightarrow Fe^{2+} + 2e^-$$

The iron (II) ions are slowly converted to oxides of iron and form a deposit as a flake of rust with the chemical formula $Fe_2O_3 \cdot nH_2O$. This hydrated oxide itself speeds up the corrosion process. While iron is oxidized, water molecules and hydrogen ions are reduced.

Rusting can be prevented in a number of ways. The iron can be coated with paint or tin, for example, to prevent contact with atmospheric moisture and oxygen. Applying a coating of zinc to iron can also help prevent corrosion, since zinc has a higher tendency to be oxidized than iron does. Zinc oxide is impervious to water. Iron coated with a protective zinc layer is called galvanized iron.

ISLANDS

This idyllic-looking tropical island is situated near the island of Oahu in the Hawaiian archipelago.

CONNECTIONS

● The **CLIMATE** of islands varies less than that of continents. This is because islands are surrounded by **OCEANS**.

● There are about 20 island arcs; most are situated in the **PACIFIC OCEAN**.

Islands typically conjure up images of peace and remoteness from life on the continents, and in some cases they may be very different from the nearest continental landmass. The nature of different islands, geographical and biological, depends on many factors, including the age of the island, its origins, and the degree to which it is isolated.

There are many different types of islands, widely varying in size and topography. Their emergence and disappearance depends on a number of factors. Many former islands are now completely submerged under the sea, and future rises in relative sea level would mean the end of many of the low-lying islands inhabited today.

Geographers distinguish between continental and oceanic islands. A continental island is one that is geographically close, and geologically related, to a continent; in fact, most continental islands were once attached to the mainland. They are separated from the nearby continent by shallow seawater, usually no more than 590 ft (180 m) deep. The British Isles and Sri Lanka are large continental islands—there are, of course, many smaller ones.

By contrast, oceanic islands are those that are geographically isolated from the continents. These islands are seamounts that rise from the floors of deep ocean basins. They are composed largely of basalt.

Oceans and continents

To get a better idea of what an island is, it is useful to review the basic geography of the world's ocean basins and continents. The land that forms Earth's crust and attaches it to the top level of the mantle is known as the lithosphere. The lithosphere under the

CORE FACTS

■ Islands are usually made up of volcanic material or limestone or both.

■ There are many different types of islands. They vary widely in size and topography.

■ Many former islands are now completely submerged, and future rises in relative sea level will mean the end of many of today's low-lying islands.

great landmasses called continents is mainly made up of granite, while the lithosphere under the oceanic basins is primarily made of basaltic rock (see IGNEOUS ROCKS). Compared to its structure underlying land, the lithosphere under the oceans is thin—only around 6 miles (9 km), compared to 34 miles (55 km) for the continents.

Over the past few decades, the theory of plate tectonics has become an accepted explanation of the structure and movement of Earth's crust (see EARTH, STRUCTURE OF; PLATE TECTONICS). The lithosphere is made up of seven major and at least a dozen minor plates of land. The plates are in constant motion, and since they lie adjacent to one another, the movement of one affects the others. What happens at the boundaries of plates explains a great deal about earthquakes, mountains, and volcanoes. It also provides explanations for island formation.

Island formation

Islands are typically made up of one or two core substances—volcanic material or limestone—or some combination of the two. Compared to continents, the geology of islands tends to be relatively simple and homogeneous. This is due to the relative "youth" of most islands, which means that there has been less time for geological evolution from their original form—ocean floor volcanoes. The climate of islands also varies less than that of the continental climate, because they are surrounded by ocean (see CLIMATE).

Advances in the understanding of global tectonics over the past few decades have led to much greater knowledge of where and how islands develop. In fact, one useful way to distinguish between types of islands is to establish whether they occur at the boundaries of plates (that is, whether they are divergent, convergent, or along transverse plate boundaries) or are well within the boundaries of a plate (that is, whether they are intraplate).

Some islands occur in areas of plate convergence. Here, plates move toward each other. As the plates collide, one typically overrides the other along a subduction zone, and islands may be produced by the resulting volcanic activity. Today, many of the islands produced in this way are still volcanically active. The multitude of islands that make up the Philippines are examples of this type of island, as are the Antilles in the Caribbean. Where two plates meet head-to-head in an area of plate convergence, crustal uplift may take place, and islands can also be formed in this way.

Another type of island is found at divergent plate boundaries. This phenomenon is associated with seafloor spreading along certain ocean ridges. Characteristics of islands located at divergent plate boundaries include volcanic activity and both lateral (sideways) and vertical tectonic movement. Iceland is an example of this type of island (see the box on page 623).

A rarer type of island is that found along transverse plate boundaries, places where one plate slides along the other with negligible convergence or

divergence. Typically, islands form in these areas only if there is some significant type of deformation of the boundary. Clipperton, in the Pacific Ocean near Mexico, is thought to be an example of this type.

For islands found within the interior of a plate, another distinction, based on the islands' distribution, can be useful. Islands occurring in a linear pattern, such as the Hawaiian islands, are thought to have resulted from intraplate volcanism, either through a fissure or at a fixed point (see below). Islands that do not form chains—isolated islands—can arise from a variety of different causes. Some scientists have speculated that they form when new plate boundaries are developing or old plate boundaries are disappearing.

This space shuttle photograph reveals the Hawaiian island chain. From the bottom right, the islands are Niihau, Kauai, Ouahu, Molokai, Lanai, Maui, Kahoolawe, and Hawaii.

WORLD'S SMALLEST AND LARGEST ISLANDS

The size of an island is determined by a number of factors, including its origin and the subsequent geological and tectonic history. The size of a given island can vary over time, depending on the surrounding sea level. For example, 18,000 years ago, the sea level in the southwest Pacific Ocean was 400 to 500 ft (120 to 150 m) lower than it is today. Many islands existed that are now completely covered by water, and islands we know today, such as New Caledonia, were much larger.

The world's smallest islands include Suwarrow in the Cook Islands, with an area of around 2 sq miles (5 km²), and Marcus Island in the Pacific Ocean, with an area of around 2⅓ sq miles (6 km²). Compare these to some of the world's largest islands: New Zealand has an area of 103,384 sq miles (270,000 km²), and Iceland and Cuba each have an area of around 40,000 sq miles (104,000 km²). The largest island is Greenland, with an area of around 840,000 sq miles (2.2 million km²).

A CLOSER LOOK

ISLAND BIOLOGY

The biological significance of islands has been recognized since the time of Charles Darwin. Darwin's studies on the islands of Galápagos off Ecuador were central to the development of his theory of species evolution.

Island biology remains curiously distinct from that of the continents, due both to the remoteness and to the smallness of the ecosystems. Small islands are unable to support a great diversity of species, since the competition for resources is intense. Of the species that are to be found, many are specially adapted to the island conditions.

Island ecosystems are often very fragile to outside disturbances. Humans have caused significant changes to island ecology by introducing foreign species. Goats, in particular, have induced significant change. During the 18th century, goats were often deliberately left to graze and multiply on islands, in order to provide food for passing ships. Over time, these animals had huge impacts on the native fauna, often leaving islands greatly denuded of vegetation.

The ocean floor

Compared to the topography of land, large areas of the ocean floor are relatively flat. They are known as abyssal plains (see OCEANS AND OCEAN ABYSSES). There are, however, local variations, such as abyssal hills and seamounts. Abyssal hills are less than 3300 ft (1000 m) high, while seamounts are steep, conical volcanoes that rise sharply from the ocean floor, sometimes reaching above the ocean surface to become islands. Flat-topped seamounts that remain submerged under water are called guyots. The upward growth of seamounts requires a sustained supply of magma from Earth's interior.

In warm, shallow waters, corals grow around seamounts to form fringing reefs. As an island subsides, the reef will appear farther away from it because the land area above the ocean surface shrinks; a barrier reef is formed. In some cases, the original island base completely subsides, leaving a ring-shaped reef, or atoll.

The most notable geological feature of the ocean floor is the system of midocean ridges and rises— long underwater volcanic mountain ranges—and trenches. In the Atlantic Ocean, for example, the mid-Atlantic ridge divides the ocean into two oceanic basins. The island of Iceland forms part of this ridge, which separates the plate carrying North America and the one carrying Europe and much of Asia. Iceland is one of the few places in the world where an oceanic ridge rises above sea level.

In addition to the mountainous ridges and rises, the ocean floor has deep trenches. These trenches are found in the Pacific Ocean and are often associated with island arcs and result from subduction of oceanic plates (see PLATE TECTONICS).

Island arcs

The chain of volcanic islands formed when tectonic plates converge at a subduction zone is called an

This aerial view of the northern part of the Palau island group in the Pacific shows the ring of coral reef just below the water surface. The ring encloses a deeper lagoon.

island arc. The name *arc* was chosen because of the curve along which the islands form.

There are about 20 island arcs located along subduction zones, and the great majority of them are found in the Pacific Ocean (see PACIFIC OCEAN), which is ringed by subduction zones. The Atlantic Ocean (see ATLANTIC OCEAN), in contrast, has passive margin boundaries.

Island arcs are characterized by both volcanic and seismic activity. The magma involved in island-arc volcanism is melted and subducted oceanic crust. The magma is caused by the heat associated with the burial of the plate in the zone of subduction, as well as the heat produced by friction between the plates.

Some island arcs are characterized by a chain of volcanically active islands and a parallel chain of volcanically inactive islands. These inactive islands are thought to result from the upthrust of the top plate, which floats on the magma beneath. An example of a double chain of islands is the Lesser Antilles in the Caribbean.

Barrier islands

Another type of island is made of sand, lies offshore, and is parallel to the coast. This type of island is called a barrier island and is usually found along most of the world's lowland coasts. Coney Island, New York, is a large-sized example of a barrier island. The long chain of islands centered at Cape Hatteras on the North Carolina coast is another, as is Padre Island, Texas, which is around 100 miles (160 km) long.

Most barrier islands were built around 5000 to 6000 years ago when the rapid rise in sea level following the end of the last glacial age began to slow down. As the sea slowly rose across the very gentle continental shelf off the southeastern United States, waves probably broke in the shallow water offshore. These caused erosion, making large piles of sand build up to form long bars. These bars gradually built up above sea level, becoming barrier islands. Then, as the sea level rose over the last several thousand years, the barrier islands moved progressively landward as the edge of the continent was slowly submerged.

A barrier island is generally made up of several ridges of sand. These relate to the successive shorelines occupied as the island formed. When a violent storm occurs, surf washes over low places in the barrier and erodes it, cutting inlets that may never disappear. Simultaneously, sediment is transported toward the land by overwash, as well as by longshore drift. The development of barrier islands depends on factors such as sediment supply, the stability of sea level, and the direction and intensity of waves and nearshore currents.

Island life

Islands can be colonized in many different ways. Large, strong flying animals, particularly migratory birds, may be caught up by strong air currents. Other creatures may use natural rafts of floating vegetation. Plants have evolved adaptations for dispersing seeds

ICELAND

Iceland is a particularly interesting example of island formation. It is one of the few places in the world where an oceanic ridge reaches the surface of the sea. In oceanic ridges, two plates diverge and the seafloor spreads. In these areas, material from the mantle is pushed upward.

In Iceland's case, this process is intensified because the island is located above one of the world's hot spots. There are approximately 40 such sites around the world, where the hot mantle forces itself to the surface. Iceland has about 200 volcanoes, and at least 30 of them have erupted since people settled on the island in the 9th and 10th centuries C.E. Altogether, there have been more than 150 eruptions. This volcanic activity is thought to be the combined result of the midoceanic ridge spreading and a hot mantle plume—a situation that produces large volumes of basaltic magma.

Iceland is home to many volcanic eruptions (see below). It is located above one of the world's hot spots, places where magma rising from the lower mantle reaches the surface.

away from the parent. Seeds that are usually dispersed by the wind, such as orchid seeds, have been known to travel up to 125 miles (200 km).

One advantage of living on an island is that there are usually few competitors. But the small size of the populations on islands makes them vulnerable to extinction by chance events, such as a small variation in the birthrate. Island populations are also vulnerable to catastrophes, such as severe storms, because they have little opportunity to escape the island and then return.

S. FENNELL

See also: CONTINENTAL SHELVES; OCEANS AND OCEAN ABYSSES; PACIFIC OCEAN; PLATE TECTONICS.

Further reading:
Nunn, P. D. *Oceanic Islands.* New York: Blackwell Publishers, 1994.
Skinner, B. J. and Porter, S. C. *The Dynamic Earth: An Introduction to Physical Geology.* New York: John Wiley & Sons, 1995.

ISOMERS

Isomers are chemical compounds with the same molecular formula but different arrangements of atoms

Milk contains lactic acid, an organic compound that exists in two isomeric forms.

The word *isomer* comes from the Greek, meaning "equal parts." Isomers are compounds that have the same molecular formula but differ in certain of their physical or chemical properties. Isomers have different properties because their molecular structures are different. The more complex the molecule, the more likely it will have numerous isomers. There are many isomers among organic compounds because of the different ways in which carbon atoms can be linked to other atoms (see CARBON).

For just one simple example, look at the compound with the molecular formula C_2H_6O. This can be either dimethyl ether (CH_3OCH_3), a gas that liquefies at -76°F (-24°C), or ethanol (CH_3CH_2OH) a liquid that boils at 172°F (78°C).

There are two main classifications of isomers: structural (or constitutional) isomers and stereoisomers. Structural isomers differ according to which constituent atoms are linked together in the compound and how these atoms are linked together. Stereoisomers, however, have the same structural constitution—the way it can be written on a two-dimensional page—but different configurations—the way the atoms are arranged in three-dimensional space.

Structural isomers

The chemicals dimethyl ether and ethanol are one example of structural isomers, because their atoms are linked together differently. Each carbon atom in dimethyl ether is connected to three hydrogen atoms and linked by one oxygen atom, and the structure looks like this:

CORE FACTS

- Isomers are compounds that have the same molecular formula but differ in certain structural, physical, or chemical properties.
- There are two main classifications of isomers: structural isomers and stereoisomers.
- Structural isomers can be subdivided into functional isomers and positional isomers.
- Stereoisomers have the same molecular formula but different orientations in space.
- Enantiomers are mirror images of each other. They have most chemical and physical properties identical.
- Diastereomers have the same functional structure but are not mirror images of their related pair of enantiomers.

$$H-\overset{\overset{\displaystyle H}{|}}{\underset{\underset{\displaystyle H}{|}}{C}}-O-\overset{\overset{\displaystyle H}{|}}{\underset{\underset{\displaystyle H}{|}}{C}}-H$$

In ethanol, however, only one carbon is linked to three hydrogen atoms; the second carbon is linked to the first carbon, two hydrogens, and the oxygen atom. The oxygen atom is linked to the sixth hydrogen atom, and the structure looks like this:

$$H-\overset{\overset{\displaystyle H}{|}}{\underset{\underset{\displaystyle H}{|}}{C}}-\overset{\overset{\displaystyle H}{|}}{\underset{\underset{\displaystyle H}{|}}{C}}-O-H$$

Structural isomers can be of different kinds. Functional isomers differ in their functional groups (see FUNCTIONAL GROUPS). In the above example, CH_3-, $-O-$, and $-CH_2OH$ are different functional groups, and the two compounds have very different chemical and physical properties.

Positional isomers have the same functional groups and differ only in their positions along a chain or in a ring. For example, hydrocarbons with four or more carbons (see HYDROCARBONS) have two or more isomers. Butane is C_4H_{10}, but these atoms can be arranged in two ways:

Pentane (C_5H_{12}) can have three isomers. An alkane containing 10 carbons could have 75 isomers, and—theoretically, at least—one containing 30 carbons could have more than 4 billion!

Alkenes and alkynes and their derivatives can also have numerous isomers, depending upon the position of the double and triple bonds. (See the section on geometric isomers, below.)

Positional isomers occur in the same way in aromatic compounds. Dinitrobenzene, for example, can exist in three isomers, depending on the position of the nitro groups relative to one another (see NITRO COMPOUNDS):

ortho-dinitrobenzene *meta*-dinitrobenzene *para*-dinitrobenzene

Positional isomers like these usually have similar physical and chemical properties. As a result, they can be very difficult to separate.

Other isomers that involve several rings of carbon atoms can differ in the number of carbon atoms in each ring. Like functional isomers, these generally have different properties.

Stereoisomers: enantiomers
The four valence bonds of carbon are not at right angles to each other, as might be suggested by the dia-

Chiral objects are non-superimposable mirror images. Although your right and left hands are mirror images of one another, their structure is not identical; only a right-hand glove will fit the right hand.

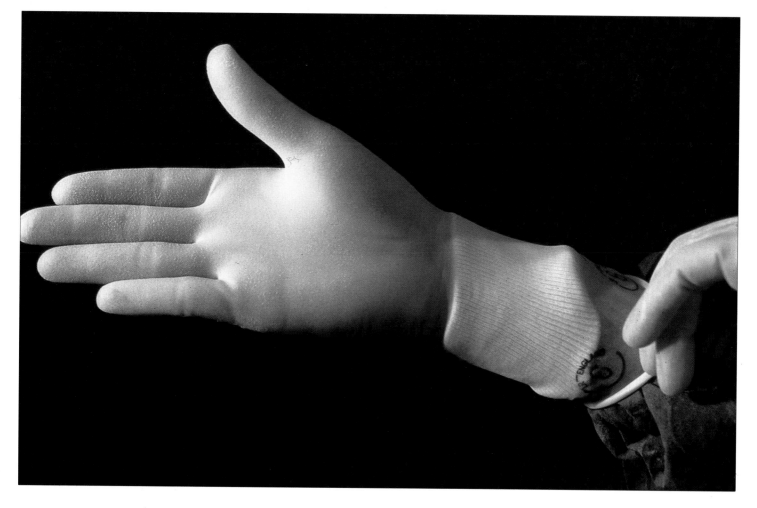

RACEMIC MIXTURES

A mixture of equal quantities of two enantiomers will tend to rotate polarized light to the right and to the left by equal amounts. Therefore, these effects cancel each other out, and the mixture will not show any optical activity at all. This type of a mixture is called racemic. Usually, when enantiomers are produced in a chemical reaction in a laboratory, they form in equal quantities and result in a racemic mixture. A challenge for chemists is to separate optical isomers from one another. The separation of one or both enantiomers from a racemate is called resolution. The most common way to separate enantiomers is with a process involving the reversible conversion of the mixture into a pair of diastereomers, by reacting it with a single pure enantiomer (diastereomers have different physical properties). This can produce diastereomeric salts, for example, which can then be separated by their different physical properties, such as solubility.

A CLOSER LOOK

Molecular models showing the two possible structures of lactic acid (carbon atoms are black, oxygen atoms are red, and hydrogen atoms are white). These structures are mirror images but cannot be superimposed on one another.

grams above, but point to the four corners of a three-dimensional shape, an equal-sided tetrahedron, with angles of 109.5 degrees. It is this fact that accounts for the occurrence of stereoisomers.

Hold your left hand in front of a mirror. The reflection looks not like your left hand, but like your right. However, a glove that would fit your right hand can never be made to fit properly on your left. No matter how you turn them, mirror images cannot be superimposed on one another.

To suggest the arrangement of atoms in space, chemists frequently adopt a dotted line and wedge convention in which bonds coming forward from a central atom are depicted as wedges or small filled triangles, while bonds extending back from the atom are shown by dotted lines (see the diagram below left).

If we look at the structure of a compound such as lactic acid, $CH_3CH(OH)COOH$, it has two molecular structures that are mirror images of each other (see the photograph below left). No matter how these structures are turned around, they cannot be made the same.

Molecules that cannot be superimposed on their mirror images are called chiral. The word comes from the Greek, meaning "handedness"—the molecules are symmetrical but different, like left and right hand gloves. Isomers of this type are called enantiomers, and the carbon atom at the center of the molecule is called a chiral center.

French chemist Louis Pasteur (1822–1895) was the first to discover the existence of enantiomers and a very important property that they have called optical activity. Of two chiral isomers, one will rotate the plane of polarized light to the right; that is, in a clockwise direction when viewed down the axis toward the oncoming light beam. This is called dextrorotatory. The other enantiomer causes the polarized beam to rotate to the left, or in a counterclockwise direction, which is called levorotatory. The two isomers are therefore designated as "d" or "l," depending on the rotation of the polarized light.

Except for their optical properties, enantiomers are very much alike. For example, they have the same melting point, boiling point, free energy, chemical properties, and X-ray diffraction pattern. Enantiomers always exist in pairs, because it is only possible for an object to have one mirror image of itself. However, most chiral substances are usually found in nature as only one enantiomer, and few natural products are represented in nature by both enantiomers. The difference between enantiomers can be dramatic—for example, one enantiomer of the amino acid leucine tastes bitter, while the other tastes sweet.

Diastereomers

It is also possible for isomers to exist that have the same structural constitution but are not mirror images of one another. This can happen when molecules have more than one chiral center.

The diagram on the left shows the four possible structures of 2,3-dichloropentane. Structures (1) and (2) are enantiomers of one another, being mirror images; so are (3) and (4). But (3) and (4) are not mirror images of (1) and (2). Stereoisomers of this sort are called diastereomers. They have similar chemical properties but different physical properties.

As the number of chiral centers increases, so does the number of stereoisomers. A compound with three chiral centers can have eight, that is, 2^3, isomers. The carbohydrate glucose (see CARBOHYDRATES), with five chiral centers, has 2^5, that is, thirty-two, isomers.

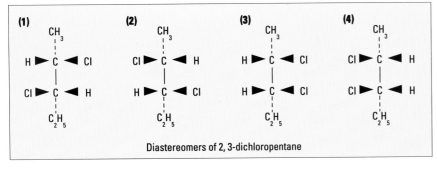

Diastereomers of 2,3-dichloropentane

USING LIGHT TO DISTINGUISH BETWEEN ENANTIOMERS

A ray of light is normally made up of light waves that vibrate at right angles to the direction of the ray, through all 360°. Many crystalline materials can block off most of these light waves, letting through only waves vibrating in a single plane (that is, at a specific angle). The light that emerges from the material is called plane-polarized light.

The crystal acts like a slit: the polarized light can only pass through a second crystal if the "slits" are aligned—that is, if the crystal is positioned at the correct angle.

Enantiomer molecules will twist the plane of polarization through an angle that is specific for that enantiomer, and a polarimeter is the

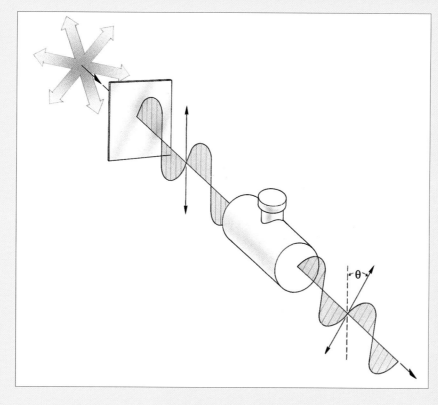

When polarized light is passed through a solution of a chiral substance, the plane of polarization is rotated either to the right or to the left. In the diagram above, the plane of polarization is being rotated to the left (observer's right).

When the tube is empty, the brightest light is seen at the eyepiece when the polarizer and analyzer are passing light in the same plane. Turning the eyepiece through 90° will dim the light to a minimum, and this effect is easier to detect. When a sample of an optically active substance is placed in the tube, the eyepiece has to be turned through an angle to find this minimum again.

If the eyepiece must be turned to the right, the substance is dextrorotatory; if it has to be turned to the left, it is levorotatory.

The degree of rotation of the polarized light will depend upon the number of molecules that it encounters in passing through the tube; that is, upon the concentration. For a substance in solution, the specific rotation is defined as the

instrument used to measure the angle through which the plane-polarized light is rotated. It consists of a light source, a polarizing lens (the polarizer), a tube that contains a sample of the material, and an eyepiece that holds a second polarizing lens, called the analyzer.

number of degrees of rotation observed in a tube 4 in (10 cm) long, containing a solution at a concentration of 1 g per ml. The light source should be restricted to a single wavelength, usually the bright yellow of the sodium flame at 5893 Å.

Geometric isomers

Atoms cannot rotate around a double bond. When a double bond exists between two carbon atoms, it fixes the atoms in place and provides the conditions for another type of stereoisomerism, called geometric, or cis-trans isomerism. The Latin word *cis* means "on this side," and *trans* means "on the other side."

The hydrocarbon 2-butene, for example, exists in two forms, depending upon the relative positions of the two methyl (CH_3) groups. In one of the structures, the two methyl groups lie on the same side of the molecule. This is the *cis* form, and the molecule is called *cis*-2-butene. In the other possible structure, the two methyl groups are positioned on opposite sides of the molecule. This is the *trans* form and is called *trans*-2-butene.

tain the same functional groups, they have similar chemical properties—but not identical: they will have different rates of reaction, for example. Their physical properties, however, will be very different. *Cis*-2-butene, for example, has a boiling point of 39.2°F (4°C) and a melting point of -218.2°F (-139°C), while the *trans* isomer boils at 33.8°F (1°C) and has a melting point of -158.8°F (-106°C).

M. RIESKE

See also: CARBOHYDRATES; CARBON; FUNCTIONAL GROUPS; HYDROCARBONS; NITRO COMPOUNDS; STEREOCHEMISTRY.

Further reading:
Amend, J. R. *General, Organic, and Biological Chemistry*. Philadelphia: Saunders College Publishing, 1990.
Hoffmann, R. *The Same and Not the Same*. New York: Columbia University Press, 1995.
Oki, M. *The Chemistry of Rotational Isomers*. New York: Springer-Verlag, 1993.
Wiegand, G. H. *Models of Matter*. St. Paul: West Publishing Company, 1995.

cis-2-butene

trans-2-butene

Geometric isomers are not mirror images of each other and so are not enantiomers. Because they con-

ISOTOPES

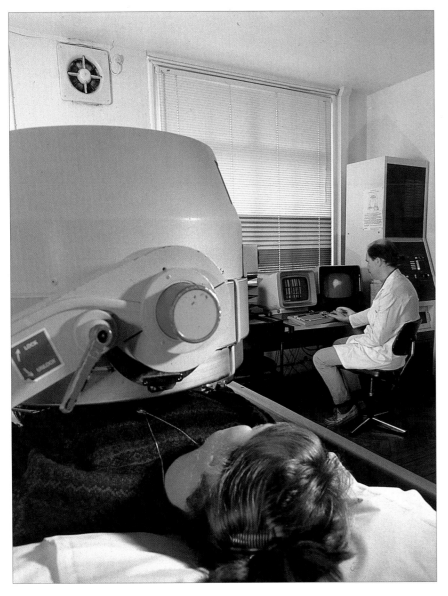

Radioactive isotopes are used to monitor and diagnose a range of disorders. The patient being scanned lies under a gamma camera, which detects the radiation.

CONNECTIONS

● The isotopes of an element have the same chemical properties, so they behave the same way in **CHEMICAL REACTIONS**.

Atoms of the same element have the same chemical properties, but their nuclear and atomic properties can differ. This is because almost all elements have more than one isotope. Atoms of different isotopes contain the same number of protons and electrons but different numbers of neutrons. Since nuclei consist of protons and neutrons, the mass of the nucleus therefore varies for different isotopes.

Individual atoms are so small that it is impossible to weigh them. It is possible, however, to measure the relative masses of the atoms in elements. Relative mass is the mass of an atom of an element compared with the mass of an atom of another element. For example, an atom of carbon is about three times the mass of a helium atom, giving a ratio of one to three.

Oxygen, for example, has three naturally occurring isotopes. The atomic number (defined as the number of protons in the atomic nucleus or the number of electrons around the atomic nucleus) of oxygen is 8, so all oxygen atoms have eight protons and

eight electrons. One isotope of oxygen has nuclei with eight neutrons, the second isotope has nine neutrons, and the third isotope has ten neutrons. For any isotope, the sum of the number of protons and the number of neutrons is called the mass number. The mass numbers of oxygen isotopes are 16, 17, and 18, and the isotopes are respectively called oxygen-16, oxygen-17, and oxygen-18. Therefore, it is the differing number of neutrons that distinguishes different isotopes of the same element. The number of neutrons in any isotope can be determined by subtracting the atomic number from the mass number.

Radioactivity

Radioactive as well as nonradioactive elements are made up of isotopes. Radioactive elements undergo the spontaneous disintegration of nuclei, with emission of nuclear particles and sometimes radiation (see RADIOACTIVITY). New nuclei are formed in the process. Uranium is an element that has radioactive isotopes, such as the isotope uranium-235 (see URANIUM). Nonradioactive isotopes include isotopes of the elements hydrogen, oxygen, and carbon.

The three most common types of radioactive decay are alpha particle decay, beta particle decay, and gamma ray emission.

In alpha particle decay, the parent nucleus—the original atom of the isotope—emits a high-speed helium-4 nucleus, composed of two protons and two neutrons, called an alpha particle. The new nucleus that is left is called the daughter nucleus, and it has a mass number that is four less than the parent nucleus and an atomic number that is two less.

In beta particle decay, the daughter nucleus formed has the same mass number as the original nucleus, but the atomic number is one greater. Beta particles are electrons that come from nuclei during beta decay. Beta decay decreases the number of neutrons by one and increases the number of protons by one. It can be thought of as the decay of a neutron to yield a proton, an electron, and an uncharged par-

CORE FACTS

■ Isotopes are atoms of the same chemical element with the same number of protons but different numbers of neutrons. Isotopes of an element have the same chemical properties.

■ Individual atoms are so small that it is impossible to weigh them directly.

■ Radioactive as well as nonradioactive elements are made up of isotopes.

■ Radioactive nuclei decay by splitting off nuclear particles and forming new nuclei.

■ Mass spectroscopy is the most accurate method of measuring relative masses of atoms.

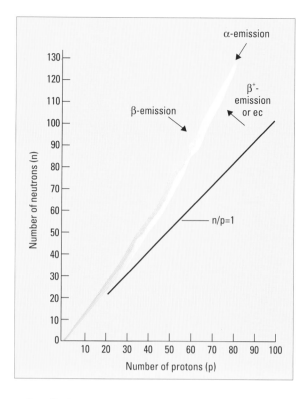

The graph to the left plots the number of neutrons against the number of protons for stable isotopes. The stable isotopes form a zone of stability. Unstable isotopes above the zone achieve stability by β-emission; unstable isotopes below the zone decay by β+-emission, or electron capture.

is one place to the right on the periodic table from the original element. Soddy was not the only chemist to discover this law—two other scientists discovered this phenomenon at about the same time, independently of Soddy.

As a result of his radioactive displacement law, Soddy found that a particular element can be produced by alpha decay or by beta decay. Either way, the element has the same atomic number but could have different atomic weights, depending on the way it was formed. From this, Soddy concluded that elements can exist in more than one form, having the same atomic numbers but different atomic weights. He decided to call these variations isotopes, which means "the same place." In 1921, he was awarded the Nobel Prize for chemistry for his discovery.

Neon isotopes

British physicists Joseph John Thomson (1856–1940) and Francis Aston (1877–1945) worked together at the Cavendish Laboratory at Cambridge University. Thomson was studying cathode rays and showed that they could be deflected not only by a magnet, but also by an electric field. In 1912, he found that when neon was used in the tube, a pair of positively deflected rays was produced. He thought this meant

ticle called an antineutrino, with the electron having enough energy to escape the atom.

Gamma ray emission occurs when the daughter nucleus produced by alpha or beta decay is left in an energetically excited state. This excited state is not the natural state of the nucleus, and it drops to a lower energy state. The nucleus emits a gamma ray, which is a form of highly penetrative radiant energy. The energy of a gamma ray is similar to, but higher than, the energy of an X ray (see X RAYS).

Isotope discovery

British chemist Frederick Soddy (1877–1956) discovered isotopes, and his research in 1896 cleared up much of the mystery surrounding radioactivity. He worked with British physicist Ernest Rutherford (1871–1937) to develop explanations for the nature of radioactivity.

Soddy and Rutherford showed that a radioactive element emits either an alpha particle or a beta particle and changes into a new element when it undergoes radioactive decay. Additional research by Soddy confirmed that helium is produced during the radioactive decay of radium (see RADIUM).

Soddy also developed the radioactive displacement law. This law describes the changes that occur when a radioactive material loses an alpha particle or a beta particle. Soddy determined that the loss of an alpha particle by an element causes that element's atomic weight to decrease by four and its atomic number to decrease by two. According to this law, therefore, alpha decay results in the formation of a new element two places to the left of the original element in the periodic table (see PERIODIC TABLE). The law further states that the loss of a beta particle has no effect on the element's atomic weight but causes the element's atomic number to increase by one. Therefore, an element produced by beta decay

ISOTOPIC FRACTIONATION

The separation of isotopes is challenging to scientists. Because isotopes of the same element have different physical properties, several methods are used to separate them. Francis Aston developed the mass spectrometer, the first device used to separate isotopes, in 1919.

During the Manhattan Project of the 1940s (the project set up by the U.S. government to develop the atomic bomb), isotopic separation was an important problem because, of the two most abundant naturally occurring isotopes of uranium, only rare uranium-235 can be used for nuclear fission (see FISSION; URANIUM). Therefore, in order for the fission bomb to be created, scientists had to find a way to separate uranium-235 from uranium-238, which is much more abundant. For this particular isotope, separation using mass spectrometers did not work properly, and this led to the development of a more efficient method devised in 1940 by John Ray Dunning (1907–1975) of the United States. He found that gaseous forms of two or more isotopes would diffuse at different rates, allowing the isotopes to be separated that way (see DIFFUSION AND OSMOSIS). The diffusion process has to occur several thousand times to obtain a gas that is significantly rich in uranium-235, but this process still became a successful and economical method for separating uranium-235 and uranium-238 isotopes.

Some additional methods of isotopic separation also exist. One scientist discovered that thermal diffusion could be used to separate uranium-235 hexafluoride from uranium-238 hexafluoride. In this, a temperature gradient was created in a liquid uranium compound and the lighter molecules containing uranium-235 were collected at the cooler part of the apparatus. This method was not used, however, after gaseous diffusion was found to be more efficient.

Isotopes of hydrogen are separated most efficiently by using electrolysis of water. Another method, distillation, also works well, but it is too expensive and no longer used commercially. A more recent method for isotopic separation uses a laser beam to excite one of the two isotopes to be separated. This is possible because of a slight difference in vibrational energy levels resulting from the difference in mass of the isotopes, so molecules containing different isotopes absorb energy at slightly different frequencies and can be separated.

A CLOSER LOOK

Frederick Soddy was a British radiochemist. He proposed the radioactive displacement law to explain what happens when a radioactive material loses an alpha or a beta particle.

nuclei of a particular element all have the same positive electric charge, but the isotopes differ in mass, each with its own mass number. Aston received the Nobel Prize for chemistry in 1922 for the invention of the mass spectrometer.

Artificial isotopes

As well as by natural processes—for example, under certain atmospheric conditions—isotopes can be made by humans. In 1934, French physicists Frédéric Joliot-Curie (1900–1958) and Irène Curie (1897–1956) bombarded aluminum atoms with alpha particles. In this bombardment process, the nucleus of the aluminum atom absorbed an alpha particle and emitted a proton. The aluminum nucleus had 13 protons and 14 neutrons, so it was called aluminum-27. When the atom took up an alpha particle, which is made up of two protons and two neutrons, and gave up a proton, it ended up with 14 protons and 16 neutrons, making it silicon-30, which occurs in nature. In some cases, however, after absorbing the alpha particle, the aluminum nucleus emitted a neutron instead of a proton, which made it phosphorus-30. This does not occur in nature; it is radioactive, with a half-life (the time in which the amount of a radioactive nuclide decays to half its original value) of less than three minutes. When it breaks down it gives off a stream of positrons, which convert protons to neutrons so that phosphorus-30 becomes stable silicon-30. This was the first experiment that produced a radioactive isotope from an ordinary stable element. The radioactivity is called artificial radioactivity, because it results from the bombardment of nuclei in a laboratory.

Uses of isotopes

Isotopes have many uses in scientific research and in medicine.

• **Scientific research:** An element's chemical properties depend on its electronic configuration, not on its atomic mass. Different isotopes of the same element therefore have the same chemical properties. Scientists use radioactive isotopes to "tag" atoms of the same element and then follow the tagged atoms through the course of a chemical reaction. This method is often used to investigate the exact pathway a reaction takes.

When chemists wanted to investigate the reaction between water and an ester, such as ethyl ethanoate, they were able to use the oxygen isotope oxygen-18. By labeling the water with oxygen-18 and using a mass spectrometer to analyze the products of the reaction, scientists found that the oxygen atom from the water attaches to ethanoic acid rather than to ethanol in the following reaction:

$$\text{H-}^{18}\text{OH} + \text{CH}_3\text{COOC}_2\text{H}_5 \rightarrow \text{CH}_3\text{CO-}^{18}\text{OH} + \text{C}_2\text{H}_5\text{-OH}$$

Isotopic labeling is widely used in biochemistry to study metabolic pathways—the chemical reactions that take place within the cells of a living organism. Carbon-14, hydrogen-3, phosphorus-32, and sul-

that either the neon was contaminated with neon hydride, or that the process produced a previously unknown element. He asked Aston to find out if either of these two assumptions was correct.

In 1919, Aston built a mass spectrometer (see the box on page 631) to separate a beam of positive rays into distinct lines. When he analyzed neon with the device, two distinct lines were produced. One line corresponded to a substance with the atomic weight 20, and the other line denoted something with the atomic weight 22. At this point, Aston concluded that neither of Thomson's predictions was correct. Instead, he decided, the two neon lines represented isotopes of neon, with masses of 20 and 22.

Aston also worked with chlorine atoms and again found two types of atoms. In the years that followed, he continued to improve his mass spectrometer, and he analyzed most of the elements that were known at that time. Ultimately, he identified 212 of the 287 stable isotopes. The use of the mass spectrometer showed that most stable elements consist of two or more stable or nonradioactive isotopes. The atomic fur-35 have all been used in biochemical research (see BIOCHEMISTRY).

• **Dating techniques:** A small amount of the carbon dioxide in the atmosphere contains the radioactive carbon isotope carbon-14. Plants take in carbon dioxide during photosynthesis, and some of the carbon-14 isotope becomes incorporated in plant tissues. Animals eat plants, and so a small amount of carbon-14 also ends up in animal tissue. Radiocarbon dating (see RADIOCARBON DATING) measures the amount of carbon-14 in organic material and can date plant and animal fragments as old as 60,000 years. Anything older has too little carbon-14 remaining to be accurately dated, but other radioisotopes, such as those of uranium, potassium, and argon, can be used for older materials. By measuring the light emitted from certain minerals when they are heated in the laboratory, the thermoluminescence method helps to determine the time that minerals such as quartz and feldspar have been exposed to the natural radioactivity within a sediment. The intensity of the light emitted indicates the length of time the substance was exposed to radiation. Older materials have been exposed to radiation for longer and emit more energy.

Oxygen isotopes are good for determining the former temperature of lake carbonates, stalactites, glacier ice, and shells preserved in deep-sea deposits. Of the two stable oxygen isotopes, one is lighter, more abundant, and evaporates more readily than the other. So atmospheric moisture, rain, and therefore ice are enriched in the lighter isotope. By measuring the ratios of the isotopes, scientists can discern periods of relatively less abundant light oxygen in the oceans. Geologists have used this method to track the advance of glaciers and to estimate the times of ice ages (see GLACIERS; ICE AGES).

• **Isotopes in medicine:** Radiotherapy uses radioactive isotopes—especially cobalt-60—to treat cancer. The isotopes produce radiation, which is targeted at the diseased tissue, where it destroys the abnormal cells.

Radioactive isotopes also have value as diagnostic tools. Radioactive substances (called radionuclides) are introduced to the body, where they are taken up in different amounts by different tissues. The tissues emit gamma rays, which are detected by a camera that rotates around the patient. Radionuclides include radioactive isotopes of elements that are found naturally in the body, such as iodine, and synthetic elements.

M. RIESKE

See also: ELEMENTS; RADIOACTIVITY; RADIOCARBON DATING.

Further reading:

Bowen, R. *Isotopes and Climates*. New York: Elsevier, 1991.
Bowman, S. *Radiocarbon Dating*. Berkeley: University of California Press, 1990.
Dickson, T. R. *Introduction to Chemistry*. 7th edition. New York: John Wiley & Sons, 1995.

MASS SPECTROSCOPY

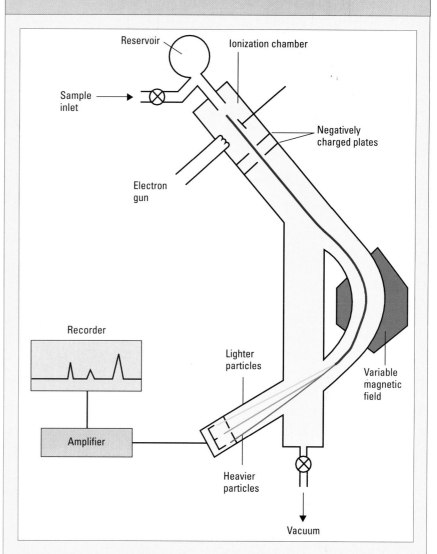

The most accurate method of measuring relative masses of atoms is mass spectroscopy. This method uses an instrument called a mass spectrometer (see the diagram above), which causes electrically charged atoms to be forced to move through a magnetic field. By measuring the curvature of the paths of atoms in a mass spectrometer, the relative masses of the atoms can be measured. The lighter the particles, the greater the deflection. The intensity of the ion beam is detected electrically, amplified, and then recorded. In the same way, when a sample of a specific element is placed in a mass spectrometer, the relative masses of its isotopes can be measured. Mass spectrometer analysis also provides a value for the percentage of each isotope in the sample.

NATURAL ISOTOPES

Radioisotope	Source
Hydrogen-13 Carbon-14	Cosmic bombardment of nitrogen-14 in the atmosphere
Strontium-90 Cesium-137	Fallout from atomic and hydrogen bombs
Radium-226 Thorium-230 Uranium-238	Uranium minerals

JET STREAMS

A jet stream is a fast-moving current of air that flows above the surface of Earth

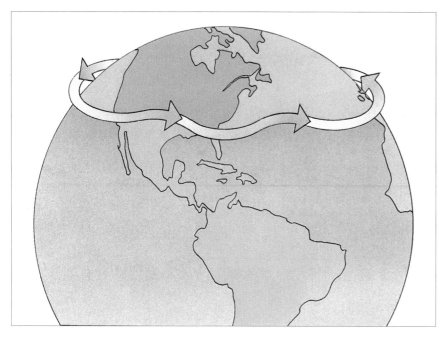

A jet stream is a current of air that travels in a wavy west-to-east direction, forming a boundary between colder air to the north and warmer air to the south.

CONNECTIONS

● Differences in **AIR PRESSURE** and **TEMPERATURE** establish **FORCES** that cause air to move from higher toward lower pressure.

● The high **WINDS** of jet streams are movements of air between **POLAR** and **TROPICAL REGIONS**.

Winds in the upper troposphere and the stratosphere blow from west to east, except near the equator. These winds are called the upper-air westerlies. The strongest cores of the westerlies are the jet streams. These fast-moving currents of air flow in the atmosphere about 6 to 9 miles (10 to 14 km) above Earth. The jet streams change their location and intensity on daily and seasonal timescales.

Jet streams were discovered during World War II by the first high-flying military aircraft. As tail winds, jet streams increase the speed and fuel efficiency of airliners. As headwinds, jet streams have the opposite effect. In fact, the bombers of World War II slowed down almost to a crawl when flying directly into jet streams. A more dangerous effect is the air turbulence created by the strong wind motions of jet streams.

Jet streams vary with the seasons. They can also affect local and global weather conditions. They can influence the location and intensity of weather fronts and areas of high and low pressure (see AIR PRESSURE AND BAROMETERS).

The high winds of jet streams are caused by the temperature and pressure differences between polar and tropical regions. The strongest jet stream winds occur in winter when these latitudinal contrasts are greatest (see WIND). At the center of the stream, winds can on occasion exceed 250 mph (400 km/h). These high speeds are found within very narrow bands about 50 miles (80 km) wide and 2 miles (3.2 km) deep in the atmosphere (see ATMOSPHERE). Jet stream winds can flow for thousands of miles, encircling Earth in a wavy, or serpentine, manner. Since they move air around Earth, jet streams are also important in the global transfer of heat.

Weather maps depict jet streams as wavy ribbons of fast-moving air. They change position constantly, moving vertically as well as horizontally. Small daily changes influence local weather. Larger, long-lasting movements from usual positions can cause droughts, heat waves, cold snaps, or floods by diverting weather systems from their normal routes (see WEATHER).

The major jet streams

There are two major jet streams in each hemisphere. The most important jet stream for air travel and weather forecasting is the polar-front jet stream, which flows in each hemisphere in a wavy but generally west-to-east manner. In the Northern Hemisphere, it generally lies between 30 degrees and 60 degrees N, at the intersection of the surface prevailing wind belts called the polar northeasterlies and the middle-altitude westerlies. The path of the polar-front jet stream varies greatly from day to day. Meteorologists observe this jet carefully and use mathematical models of the atmosphere to predict its movements for one to four days in advance.

The second major jet stream, the subtropical jet, flows from west to east, usually roughly southwest to northeast, between 2 degrees and 50 degrees latitude in each hemisphere. The subtropical jet occurs roughly at the interface between the surface prevailing wind belt called the trades and the westerlies of the mid-latitudes, and is accompanied by sporadic periods of strong convection and heavy rain showers.

Jet streams change with the seasons, largely in response to changes in north-south temperature contrasts between the summer and winter. They generally weaken in the summer and move farther north. The subtropical jet all but disappears, except for a segment over the Mediterranean that blows from west to east at lower speeds.

There are several other minor and short-lived jet streams in the atmosphere. One of these is the equatorial jet stream that flows from east to west and occurs only over Southeast Asia and Africa in the summer. This jet stream is believed to be related to the warming of the air over large elevated landmasses. This jet stream determines the coming and duration of Indian and African summer monsoons (see MONSOON), so it is vital to agriculture. Another jet stream, the polar night jet, occurs in the atmospheric layer above 9 miles (14 km), moves in an easterly direction, and occurs only in the winter.

P. WEIS-TAYLOR

See also: ATMOSPHERE; CLIMATE; METEOROLOGY; WEATHER; WIND.

Further reading:

Allaby, M. *Air: The Nature of Atmosphere and the Climate.* New York: Facts On File, 1992.
Encyclopedia of Climatology. New York: Van Nostrand Reinhold Company, 1992.

JUPITER

Jupiter is the largest and most massive planet in the Solar System

CORE FACTS

- Jupiter is the largest, most massive planet in our Solar System.
- Jupiter's Great Red Spot is a storm that has lasted for centuries.
- Most of Jupiter, including its atmosphere, is made up of hydrogen and helium.
- Jupiter generates an intense magnetic field.

JUPITER

- Distance from the Sun: 484 million miles (778 million km)
- Orbital period: 11.86 Earth years
- Rotation period: 9 h 55 m 29 s (but differential)
- Equatorial diameter: 88,700 miles (142,700 km)
- Polar diameter: 82,950 miles (133,500 km)
- Mass: 2.1×10^{24} tons (1.88×10^{24} tonnes)
- Average density: 83.28 lb/cu ft (1.34 g/cm³)

Jupiter, the fifth planet from the Sun, is named for the god who the Romans thought ruled the sky and the weather. Jupiter, also known as Jove, was one of the most powerful deities of ancient Rome.

Jupiter, like all planets, was formed during star birth. In this process, according to one theory, an interstellar cloud of gas and dust began to condense, eventually forming a solar nebula (SEE NEBULAS). As the mass continued to contract under its own gravitation, planets formed. Planets can be rocky and metallic bodies, icy bodies, or large, gaseous bodies. The rocky and metallic bodies are called terrestrial planets. These include Mercury, Venus, Earth, and Mars. Large gaseous bodies are called Jovian planets and include Jupiter, Saturn, Uranus, and Neptune. Jupiter is the largest, most massive planet in our Solar System; it has twice the mass of all the other planets combined. Although it is very large, Jupiter is made up of low-density materials.

The ancients could see Jupiter; it is one of the brightest objects in the sky. But the colorful clouds that swirl in Jupiter's atmosphere were not seen

This color-enhanced image of Jupiter shows the cloud-enveloped planet and its Great Red Spot, which is thought to be a gigantic storm.

CONNECTIONS

- Jupiter radiates thermal **ENERGY** into **SPACE** as a result of a contraction that converts gravitational energy into **HEAT** energy.

FINGERPRINT OF THE COSMOS?

There are two major theories about how Jupiter formed. One theory is that as the primordial cloud was coalescing, an initial core of icy rock could have pulled in the Jovian atmosphere. In fact, satellite gravity measurements indicate that Jupiter may have a rocky core. The second theory is that Jupiter might have been made the same way as our Sun (see SUN): elements from the primordial Solar System nebula could have coalesced, forming a starlike planet.

In either event, astronomers theorize that Jupiter and the other Jovian planets are primitive. They have evolved little and, in Jupiter's case at least, massive gravity has kept primordial elements in the planet's atmosphere. Thus, the composition of Jupiter should be similar to the composition of the original gas and dust cloud from which our Solar System was originally formed some 4.5 billion years ago (see SOLAR SYSTEM).

A CLOSER LOOK

until telescopes became available. The atmosphere of Jupiter is dominated by these cloudy regions, called zones, which are separated by darker areas, called belts. Astronomers think that the zones are regions of high pressure, in which atmospheric gases are rising. Belts are thought to be regions of low pressure and descending atmosphere.

The bands of color on Jupiter can persist for decades. The colors seem to correlate with depth in the Jovian atmosphere: for example, clouds at the top of the atmosphere are red. The temperature of the visible cloud tops is estimated to be -236°F (-149°C), consistent with the assumption that these clouds are largely made up of ice crystals of ammonia. As atmos-

These cloud patterns on Jupiter were recorded by the Voyager 1 *space probe.*

pheric height decreases, cloud features change from white to brown to deepest blue (see the diagram on page 635). Scientists who model the atmosphere of Jupiter suggest that these lower layers of clouds result from the reactions of minor constituents of the atmosphere, particularly sulfur, with the major constituents, ammonia and hydrogen. The lower cloud levels are made of ammonium polysulfides and hydrogen polysulfides. These provide the brown color. Possibly water ice is present also.

Days go by very quickly on Jupiter. The giant planet rotates so fast that its spherical shape is distorted, making the poles appear flat. In fact, the rotation of Jupiter's atmosphere varies with latitude; for example, at the equator, it rotates once every nine hours and fifty minutes, but at the poles it takes five minutes longer. This is called differential rotation.

The Great Red Spot

A storm that has been going on in Jupiter's atmosphere for centuries, the Great Red Spot is located in the south temperate region of the planet at 75 degrees longitude. To an observer on Earth, the Great Red Spot changes in size and brightness and rotates counterclockwise once every seven days or so. Cloud tops in the Great Red Spot are regions of high pressure that are colder than and above the atmosphere.

The Red Spot on Jupiter has been known since 1664, when English scientist Robert Hooke (1635–1703) first noted it. This spot occasionally darkens but then reblooms in fresh color. It is about 25,000 miles (40,000 km) long and 9000 miles (14,000 km) wide.

Jupiter's atmosphere

Although observers on Earth identified ammonia, methane, and water in Jupiter's atmosphere, most of the planet is made up of hydrogen (82 percent by mass) and helium (12 percent by mass). Thus, most of Jupiter's upper atmosphere is made up of hydrogen and helium gases. Deeper in the atmosphere, the increasing pressure causes these gases to liquefy. Lower still in the Jovian atmosphere, liquid hydrogen molecules separate into their atoms. This form of hydrogen is known as liquid metallic hydrogen because it conducts electricity. This transformation occurs at a pressure 2 million times that of Earth's atmosphere. No one knows if Jupiter has a solid core. If one exists, it is probably only 18,000 miles (30,000 km) in diameter and made up of heavy elements.

Early Galileo probe results

The first measurements of Jupiter's atmosphere were made on December 7, 1995, when the Galileo probe plummeted into Jupiter's atmosphere. The probe's findings indicate that Jupiter is a complex world.

Galileo found some surprises beneath the high clouds, which had long been visible from Earth and were more recently surveyed by the Hubble Space Telescope (see HUBBLE SPACE TELESCOPE) and by Voyager (see VOYAGER PROBES) and Pioneer spacecraft (see PIONEER). First, the atmosphere was drier than predicted. But astronomers think this may be because

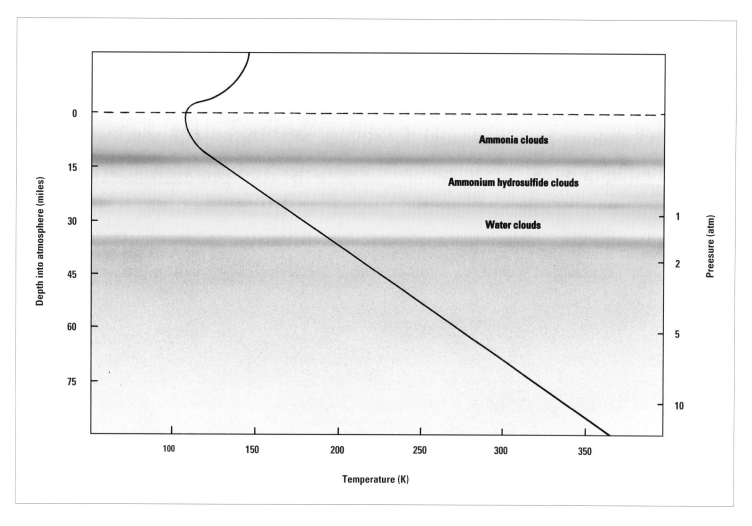

the probe entered a hot spot. Given that Galileo's measurements may have been made in an unusual region, the results may not be typical of the entire planet. For example, the probe did not find the dense clouds expected. The only well-defined cloud layer appears to be an ammonium hydrosulfide cloud layer that was previously thought to be the uppermost cloud layer on Jupiter.

Galileo also reported that several key elements and compounds appear to be less abundant than expected. These include water vapor (an indication of the amount of oxygen present), carbon in the form of methane gas, and sulfur in the form of hydrogen sulfide. Neon and other noble gases were also less abundant than predicted. However, nitrogen, in the form of ammonia, appears to be more abundant.

In the lower part of Galileo's descent, before it burned up at a temperature of 305°F (152°C) and a pressure about 23 times that of Earth's atmosphere, the probe reported temperatures close to those expected. It found the atmosphere to be well mixed. Strong winds persisted at essentially all altitudes measured. Winds below the clouds were 330 mph (540 km/h) and are the same as those the Hubble Space Telescope measured at visible cloud tops. If the winds are consistent as altitude decreases, then they cannot be produced by heating due to sunlight or condensation of water vapor, as winds are generated on Earth. Rather, they must be affected by heat being produced in Jupiter's deep interior.

It has long been known that Jupiter emits much more thermal energy than it can absorb from the Sun. Since sunlight cannot be the only source of heat, other sources of heat remain to be explained. Most scientists believe that Jupiter is still contracting and that this release of gravitational energy is a major source of heat.

Lightning on Jupiter occurs much less frequently than on Earth. The Galileo probe detected lightning discharges at radio frequencies. This may explain the sporadic radio noise that astronomers have long studied. Jupiter also emits energy in the microwave and infrared regions.

The Jovian magnetosphere

A surprise from the Galileo probe observations was a previously unknown intense belt of radiation between Jupiter's ring (a faint ring of dust and rock fragments in Jupiter's equatorial plane) and the top of the planet's atmosphere. Radiation in this belt is 10 times stronger than that in Earth's Van Allen belts (belts of intense radiation surrounding Earth, consisting of high energy charged particles trapped in Earth's magnetic field; see VAN ALLEN BELTS).

Charged particles, such as the ions emitted by the Sun, are deflected by a magnetic field. The magnetosphere is the region of space surrounding a planet, and its magnetic field is essentially the same as that of the planet. It defines the region that is shielded from the solar ions. Interactions between

JOVIAN MOONS

Jupiter has at least 16 moons, 4 of which were discovered by Italian astronomer Galileo Galilei (1564–1642) in 1610. These moons—Io, Europa, Ganymede, and Callisto—are therefore called the Galilean satellites. They are bright and larger than our Moon.

Io is the most volcanically active body known in our Solar System. It has a thin atmosphere of sulfur dioxide, probably as a result of the near-constant volcanic eruptions. Volcanoes and lakes of molten sulfur are concentrated in the equatorial regions. There are no impact craters. Io orbits at a distance equivalent to three diameters of Jupiter away from the planet. It probably has a silicate mantle and crust, over a core of molten and solid sulfur and sulfur dioxide. Material sputtered off the surface of Io forms a doughnut-shaped region of dense plasma (gaslike mixture of charged particles) around Jupiter. This is called a torus.

Europa is rocky. Voyager photos show massive cracks that run thousands of miles and gape as much as 25 miles (40 km) wide on the surface of Europa. But they are not very deep. This may indicate a thin, icy coating on an otherwise rocky but smooth-surfaced moon.

Ganymede and Callisto both probably have liquid cores and icy crusts. Ganymede is the largest Jovian moon and, in fact, the largest known moon in our Solar System. It is 3270 miles in diameter (5270 km) and, unlike that of Europa, its surface is both grooved and cratered. The craters are probably from impacts some 4 billion years ago. Callisto, the farthest Galilean moon from Jupiter, is the most heavily cratered body in the Solar System. The craters range in size but are all relatively shallow, indicating that the surface is icy. In fact, a region on Callisto named Valhalla looks like a frozen water surface disturbed by a thrown pebble. In this case, the "pebble" was large enough to create a 370-mile- (600-km-) wide crater and melt the surrounding icy surface. As the freshly melted water flowed out from the impact, it quickly refroze, preserving its wave crests around the crater.

The other Jovian moons may be asteroids captured into orbit around Jupiter, or they could be remnants of other larger bodies. Amalthea, the innermost moon, looks like an asteroid in that it is not spherical. Its diameter is between 95 and 170 miles (155 and 270 km). It is one of four moons inside the orbit of Io and has been known for about a century. The other three were unknown until the Voyager mission (see VOYAGER PROBES). Of the other eight named satellites beyond Io, four are in highly inclined orbits, and the four farthest from Jupiter are in retrograde orbits—that is, these moons orbit Jupiter in the direction opposite that of the rotation of the other planets and satellites. A moon in a retrograde orbit is thought to have been captured by the planet long after planetary formation, since moons that formed along with the planet are believed to orbit in the same direction as the planetary rotation.

These named moons are, in a sense, not the only Jovian moons. There are millions of other tiny moons that form the ring of Jupiter. These "moons" are really the size of dust fragments, perhaps as small as 10 micrometers.

the magnetic field and the solar ions draw the magnetosphere into a teardrop shape, with a long tail facing away from the Sun. This teardrop shape is similar to the region of undisturbed water that would be seen behind a spherical obstacle placed in a flowing stream. The planet's magnetic field acts as a barrier, obstructing the flow of ions from the Sun. Jupiter's magnetic field is so strong that its magnetosphere occupies a huge volume of space. The Voyager spacecraft, which flew past Jupiter, detected the tail of this teardrop out to 8700 Jovian radii, or almost 19 million miles (30 million km), behind the planet.

J. DENNETT

See also: ASTRONOMY; SOLAR SYSTEM; SPACE; SUN.

Further reading:
Beebe, R. *Jupiter: The Giant Planet*. Washington, D.C.: Smithsonian Institution Press, 1994.
Kaufmann, W. *Universe*. New York: W. H. Freeman and Company, 1994.

JURASSIC PERIOD

The Jurassic period is the geological period that lasted from 208 to 144 million years ago

The Jurassic period forms the middle part of the Mesozoic era, coming after the Triassic period and before the Cretaceous period. It lasted from about 208 million to about 144 million years ago. This period in geological history takes its name from the Jura Mountains along the border between France and Switzerland. The name was coined in 1799 by German scientist Alexander von Humboldt (1769–1859) after the sequence of rock strata in these mountains, which were studied by French geologist Alexandre Brongniart (1770–1847).

Geologists subdivide the Jurassic into upper, middle, and lower series or epochs. The early Jurassic

lasted from the end of the Triassic period until about 178 million years ago, the middle Jurassic from about 178 until about 157 million years ago, and the late Jurassic from 157 million years ago until the start of the Cretaceous period

Landmasses and movements

Throughout the Mesozoic era, the continents occupied a more southerly position than they do now. At the start of the Jurassic period, there was a single southern landmass that is called the supercontinent of Gondwana (see GONDWANA), and a single northern supercontinent called Laurasia. The process of continental drift gradually changed the distribution of land and sea into something more like what exists today.

In the middle Jurassic, North and South America became separated, and the central part of the Atlantic Ocean developed. In this period, shallower seas began to form between Africa and Laurasia, as well as between North America and Asia.

By the late Jurassic, Gondwana began to split up into the continents we now recognize as South America, Africa, India, Antarctica, and Australia, although the latter two continents remained connected at this stage. India split away from the southern continent and began drifting north toward Asia.

This still left a more or less continuous landmass stretching from the North to South Poles and a huge ocean covering much of the globe. An inlet of this

This artist's impression of the Jurassic landscape shows the dominant terrestrial animals at that time (the dinosaurs) and fauna such as conifers and ferns.

CORE FACTS

- The Jurassic period began about 208 million years ago, at the end of the Triassic, and lasted until 144 million years ago, at the beginning of the Cretaceous.
- During the Jurassic period, the continents we now know as South America, Africa, India, Antarctica, and Australia began to develop.
- Dinosaurs were the most successful land vertebrates throughout the Jurassic, and the first birds evolved during this period.
- The mollusks, called ammonites, were widespread in the Jurassic ocean, and their fossils are important in dating the ages of different rocks.

CONNECTIONS

- There is no evidence of **ICE AGES** occurring in the Jurassic.

- **FOSSILS** of plants and animals have enabled scientists to re-create an imagined Jurassic scene.

- In the late Jurassic, the **CONTINENTS** began to form.

These layers of Jurassic rock, near Lulworth, Dorset, England, were originally deposited as horizontal beds. It was the collision between African and European crustal plates that folded and tilted them.

ocean, Tethys Sea, separated the fracturing Gondwana in the south from the northern continents of the developing Europe and Asia.

Marine sediments accumulated in areas of Earth's crust occupied by seas. Tethys Sea, for example, covered an area from the position of the present Mediterranean to the Himalayas, and as far as Indonesia. Jurassic deposits include graywackes (fine to coarse, poorly sorted fragments that are firmly cemented), shales, and siliceous sediments found, for example, around the Pacific basin. Marine Jurassic rocks also occur in the southern United States, eastern Mexico, and northern Central America.

Climate

In general, climates in the Jurassic period seem to have been rather mild and moist, as well as warmer and more uniform than today. No evidence of an ice age has been found. The year was probably divided into seasons; growth rings in fossil trees provide evidence of this.

Climate is greatly influenced by the extent of oceans, since water retains the Sun's heat better than rock, leading to warmer, less extreme weather patterns, especially on land close to the oceans. During the Jurassic period, warm, dry climates were found farther from the tropics than they are today. The western seaboards of North and South America, in particular, had more extensive deserts.

Scientists have been able to detect changes in types of fossil plants corresponding to different latitudes at the time. This has shown that, in the

Northern Hemisphere at least, there was a series of zones from north to south. These indicate differences in climate. In Gondwana, however, this zoning is less clear, with the southern Jurassic flora being much more widespread and uniform.

Some of these Jurassic plant fossils resulted in deposits of coal in Scotland and Greenland, for example. In addition, Jurassic rocks have also yielded moderate amounts of mineral resources, including iron ore and petroleum in western Europe. Petroleum is also extracted from Jurassic rocks in North America, Morocco, and Saudi Arabia.

Plants

Finds of fossil plants and animals from around the world have enabled scientists to re-create an imagined Jurassic scene. With the fossil evidence, we can be fairly confident about some of the groups of animals and plants that existed.

The main groups of Jurassic plants were the club mosses, horsetails, and ferns (the simplest vascular plants, restricted to moist or wet habitats; sometimes known collectively as the pteridophytes), and the seed ferns (now extinct), ginkgos, cycads, and conifers (the simplest seed plants; known collectively as the gymnosperms). These plants were fairly uniformly distributed throughout the world at that time, and their occurrence and distribution have helped scientists to map the breakup of the continents.

The dominant terrestrial vegetation consisted of gymnosperm plants (plants with seeds not enclosed in an ovary, such as conifers), which were much more varied and abundant in the Jurassic period than they are today. This group includes the cycads (palmlike tropical plants, still represented today) and the related bennettites or cycadeoids (now extinct), as well as the conifers and ginkgos (large trees with small fan-shaped leaves).

Ferns probably formed the ground cover in many of the coniferous forests of the Jurassic, and some of the species growing then were closely related to modern types. Horsetails would also have looked familiar; these grew on damp ground, such as in the swampy soil around river deltas. Some ferns, called tree ferns, had tall trunks and reached a height of 13 ft (4 m) or more.

Cycads and bennettites, which look superficially like tree ferns, had woody trunks and grew like trees up to several yards tall. They reproduced by spore-producing cones. A typical Jurassic bennettite was *Williamsonia*, which grew up to 3 ft (1 m) and had palmlike leaves and star-shaped flowers.

Many of the modern conifer families, such as yews, monkey puzzles, cypresses and junipers, swamp cypresses, and redwoods, had evolved by the Jurassic. Ginkgos were much more widespread, especially in northern Asia, although today this group is represented by just a single species. Ginkgos grow to about 120 ft (35 m) and have very characteristic fan-shaped, notched leaves.

The seed ferns are an interesting group of plants. They are considered by some botanists to be the ancestors of the flowering plants (angiosperms). Seed ferns or pteridosperms were widespread in the Jurassic. They had fernlike foliage but reproduced using seeds. They mainly grew in warm, swampy sites, and some reached a height of more than 17 ft (5 m). A common seed fern in the Jurassic period was the genus *Pachypteris*, which grew to some 7 ft (2 m) and had lobed fronds.

Animals

Little is known about the land animals of the early and middle Jurassic. The mammal-like reptiles became extinct by the early Jurassic and were replaced by other reptiles, including the dinosaurs. The dinosaurs were the successful large vertebrates on land throughout the Jurassic.

By the late Jurassic, the land ecosystems were dominated by the carnosaur group of dinosaurs, such as *Allosaurus* in North America. The carnosaurs were large, with thick tails, short necks, and small front limbs. They had large, sharp teeth and were active predators of smaller reptiles and of some of the large, though less agile, herbivores.

Dominant among the herbivores were the huge sauropods such as *Diplodocus*, along with *Stegosaurus*. The long-necked sauropods may have reached up into the branches of tall trees to browse.

Some late Jurassic deposits, such as those in the Dinosaur National Monument in Utah, have revealed trackways of herbivorous dinosaurs such as *Apatosaurus*, *Barosaurus*, *Camarasaurus,* and *Diplodocus*. These probably grazed across the wide plains, gain-

ing protection from their size and herding instincts, like elephants do today. Other herbivores, notably *Stegosaurus*, were heavily armored, protecting them from all but the largest carnosaurs. *Camptosaurus* was a relative of *Iguanodon* and used its long snout to feed on the lower foliage of trees. *Dryosaurus* and *Ornitholestes* were smaller dinosaurs that could run quickly, often escaping their predators. In Utah, the dominant carnivores were dinosaurs such as *Allosaurus* and *Ceratosaurus*, both formidable predators with large jaws armed with long, sharp teeth.

Although the land animals would have seemed strange to us, some much more familiar animals inhabited the water. These included amphibians, such as frogs, and reptiles, such as turtles and crocodiles. These closely resembled the modern forms. Reptiles, such as the large ichthyosaurs and plesiosaurs, also lived in the seas during the Jurassic. The main groups of fish were the sharks and ray-finned fish.

Mammals first appeared in the late Triassic, and during the Jurassic they diversified into at least five orders. Nevertheless, most Jurassic mammals were small, nocturnal insect-eaters and not a significant group at that time.

Even the air had its reptile inhabitants, such as the pterodactyls and other pterosaurs (flying reptiles). However, another aerial development in the Jurassic was to have a huge and continuing impact on Earth's ecosystems. This was the evolution of birds.

The most famous early bird is *Archaeopteryx* (from the late Jurassic, around 150 million years ago), whose fossils were first discovered in 1861. For many years this was the oldest known bird, and it formed a plausible link with the reptiles. Recently, three species of fossil birds have been discovered in

This large flesh-eating dinosaur, Allosaurus, *was alive in the Jurassic period.*

THE RISE OF THE DINOSAURS

Dinosaur (which means "terrifying lizard" in Greek) is the name given to various kinds of large reptiles of the Mesozoic era. The term was first used in 1841 at a scientific meeting in England, referring to some large fossil bones unearthed in southern England.

The dinosaurs first came to prominence during the middle of the Triassic period, and they were the dominant animals all the way through the late Triassic and the whole Jurassic period, until the end of the Cretaceous period, some 165 million years later.

Dinosaurs quickly diversified and exploited the many new habitats opening up as the land plants covered the fertile, often swampy, habitats. Some scientists point to the modifications of dinosaurs' limbs and hips to explain their success; their improved stance gave them the flexibility they needed to adapt to new habitats, where they were relatively free from competition.

The various kinds of dinosaurs are classified in two basic categories—the orders Saurischia (lizard-hipped) and Ornithischia (bird-hipped)—within the reptile group archosaurs.

The largest dinosaurs, and the largest land animals ever, were the sauropods, such as *Brachiosaurus* (late Jurassic, East Africa); and *Apatosaurus*, *Barosaurus*, *Camarasaurus*, and *Diplodocus* (from the late Jurassic of North America). *Diplodocus* was one of the longest, at about 90 ft (27 m), although *Ultrasaurus* may have reached 100 ft (30 m). Hollow chambers inside their vertebrae helped reduce the weight of these giant herbivores.

The sauropods were the dominant herbivores throughout the Jurassic, but they seem only to have been of minor importance during the Cretaceous.

The stegosaurs, or plated dinosaurs, are represented by relatively few fossils, mainly from Jurassic rocks in North America, East Africa, and Europe. They include the *Kentrosaurus*, as well as the better-known *Stegosaurus*. Their bony plates are thought to have been part of a temperature control system. These plates could be angled toward the Sun to heat the blood, like solar panels, or alternatively give a cooling effect in a breeze or in cold air. Protection against predators came mainly from the bony spikes on the tail.

Allosaurus was a typical meat-eating dinosaur of the late Jurassic period, about 150 million years ago. It used the huge claws on its forelimbs, and its powerful jaws, to attack its prey.

Some of the most famous dinosaur fossils have come from the Morrison Formation, a sequence of freshwater sediments deposited in Utah, Colorado, Wyoming, and Montana. Examples include the National Dinosaur Monument and the Cleveland-Lloyd Dinosaur Quarry, both in Utah, and Garden Park in Colorado.

Cerro Condor is one of several dinosaur sites in Argentina. It is perhaps best known for its skeletons of the large sauropod *Patagosaurus* and the predatory carnosaur *Piatnitzkysaurus*. The latter was like a smaller version of *Allosaurus*.

In Europe, a slate bed near Stonesfield in England has revealed a number of important dinosaurs, including the carnivorous *Megalosaurus*, whose fossilized footprints have also been found in southern England. *Compsognathus* is one of the smallest dinosaurs known, measuring only about 40 in (1 m). Its remains have been found in Germany and France. It probably ran around in a henlike fashion, feeding on insects and lizards that it caught with its sharp teeth. *Camptosaurus* was similar to *Iguanodon* (a Cretaceous dinosaur) but somewhat smaller. It was very widespread in the Jurassic period, its remains having been found in Portugal and North America.

Tendaguru, in Tanzania, Africa, was first excavated by German scientists between 1909 and 1912. Dinosaurs uncovered there include *Brachiosaurus*, *Barosaurus*, and *Kentrosaurus*.

Many dinosaurs have been discovered in China, among them *Mamenchisaurus*, a sauropod with a huge neck some 50 ft (15 m) long, and *Tuojiangosaurus*, a stegosaur with double rows of bony plates along its back.

Other important fossil discoveries dating from the Jurassic period include those in the Solnhofen limestone in Bavaria, and in the shale deposits of Württemberg, both in Germany. Solnhofen is famous as the site where fossils of the early bird, *Archaeopteryx*, have been found. In the Württemberg shales, among others, there are fossils of ichthyosaurs, clearly showing their fins.

These fossilized ammonites contain brilliant-colored crystallized minerals. Living ammonites were marine mollusks.

Liaoning Province, China, and these all date from the late Jurassic. One of these had a beak similar to those of modern birds, while the others had features such as an enlarged sternum, very characteristic of birds and indicative of active flight. These Chinese birds were found in a layer of Jurassic mudstone with bands of volcanic ash and have been dated as slightly younger than *Archaeopteryx*. These exciting finds indicate that birds probably appeared some 60 million years earlier than previously reckoned, and that they were already quite diverse by the late Jurassic.

Many invertebrate groups were represented in the Jurassic period, notably ammonites, bivalve mollusks, and arthropods (most prominently the insects). Most insects of Jurassic age can be assigned to modern orders. These include most of the wingless types, as well as winged groups such as beetles, bugs, cockroaches, and grasshoppers. More advanced insects such as flies and hymenopterans (early ants and wasps) also evolved in the Jurassic.

Life was clearly abundant in the Jurassic, and the opportunities offered by the developing land vegetation were fully exploited, particularly by the insects and the dinosaurs. In the Jurassic seas, ammonites and bivalve mollusks were widespread. Many of these are preserved in fossils. Ammonites appear to have come close to extinction at the close of the Triassic but flourished in the Jurassic.

Dating by ammonites

Ammonites are cephalopod mollusks, belonging to the class that contains squid. Ammonites form part of a larger group called the ammonoids. Paleontologists tend to reserve the term *ammonite* for the later ammonoids with finely marked shells. These marks are called suture patterns, and they are the line of intersection between the shell and the wall of an internal chamber.

Ammonites became extinct at the end of the Cretaceous period, about the same time as the dinosaurs disappeared. The closest living relative of the ammonites is the pearly *Nautilus*, still found in the South Pacific.

We can only speculate on how ammonites lived, but their structure indicates that most of them inhabited the open seas. Many ammonites were free-swimming, using a method of jet propulsion, and others drifted or floated in surface currents. Some may have spent their adult lives on the ocean floor.

Ammonites first appeared as fossils early in the Devonian period (408 to 360 million years ago). After this, they seem to have diversified very rapidly, before declining in the late Devonian. They declined toward the end of the Triassic (208 million years ago) but came to prominence again in the Jurassic. Their rapid decline and ultimate extinction at the end of the Cretaceous may have partly been caused by the rise to prominence of speedy carnivorous ocean predators, such as marine reptiles and bony fish.

Ammonites are particularly important for geologists because they can be used to date rocks. Similar associations of ammonite species indicate rocks of comparable age, and in this way rock strata from different regions may be correlated. Ammonites are found preserved in many types of sedimentary rock, over a period spanning some 325 million years.

M. WALTERS

See also: CRETACEOUS PERIOD; GEOLOGIC TIMESCALE; TRIASSIC PERIOD.

Further reading:
Levin, H. L. *The Earth Through Time*. 5th edition. Fort Worth: Harcourt, Brace, Jovanovich, 1996.
Wicander, R. and Monroe, J. S. *Historical Geology*. 2nd edition. St. Paul: West Publishing Company, 1992.

MESOZOIC OROGENY

Orogeny is the term used to describe mountain-building activity in Earth's crust. In the Jurassic period, this activity increased quite dramatically over that in the previous period, the Triassic, and continued at an intense level in the Cretaceous period, which followed. Orogenic processes were taking place in many parts of the world at that time, notably in the Caucasus, in northeast Siberia, across much of China, and in New Zealand.

In western North America, the Nevadan Orogeny produced the Sierra Nevada. These mountains formed as the eastern Pacific tectonic plate moved underneath the American plate, thus exerting enormous forces. These forces acted to deform the crust in the area that is now known as western Nevada. It also affected a belt of land to the north, as far as British Columbia and Alaska, creating extensive fold-thrust mountain systems (see MOUNTAINS).

A CLOSER LOOK

KINEMATICS

Kinematics is the mathematical description of movement

A huge acceleration is needed to propel the space shuttle Columbia *upward.*

CONNECTIONS

● **GRAVITY** makes objects accelerate in a downward direction.

● **CENTRIFUGAL AND CENTRIPETAL FORCES** act on a body when it is following a circular path.

Motion is a fundamental condition of nature. From the smallest subatomic particle to the largest galaxy—every object moves. Kinematics is a branch of kinetics that provides a way to analyze motion without taking into consideration aspects of mass and force.

Speed and velocity

A body's motion is its change in position relative to a fixed frame of reference, and the rate of change in position is called speed. Average speed (s) is therefore a function of time (t) and distance (d), such that

$$s = \frac{d}{t}, \text{ or } d = st$$

A car covering 360 miles (580 km) in 6 hours averages 60 miles per hour (60 mph), or 100 kilometers per hour (100 km/h), that is, one mile (1.6 km) per minute, or 88 ft per second (26.8 m/s). Remember this is only the average speed over the 6-hour journey: for part of the journey, the car may be standing at traffic lights; at other times it may be traveling at 70 mph (113 km/h) or more.

However, motion only occurs with direction, and direction can change. To indicate speed in a given direction, scientists use the term *velocity*. This means a truck traveling at 60 mph (100 km/h) north has a different velocity from an identical truck traveling at 60 mph (100 km/h) northwest.

Relative velocity

A frame of reference is important when considering how objects move in relation to one another. When we say that a train is traveling at 60 mph (100 km/h), the frame of reference is probably the track on which the train is moving. It serves as the common reference point against which to measure velocity. Yet the track is not the only possible frame of reference. A train may move along at 60 mph (100 km/h) in relation to the rails, but in relation to another train moving at 40 mph (64 km/h) in the same direction, the train is only traveling at 20 mph (32 km/h). The change in velocity depends upon the frame of reference used, and it is called relative velocity (see RELATIVITY).

Acceleration

When the velocity of an object is constant, it moves in one direction with an unchanging speed. When acted on by forces, bodies speed up or slow down. The rate of these changes is called acceleration. Acceleration—that is, speeding up—or deceleration (negative acceleration) occurs, for instance, when a car goes from a standstill to cruising speed or slows down to a stop. If we know the initial velocity (v_0) and the velocity at time t (v_t), average acceleration (a) can be calculated:

$$\frac{\text{average}}{\text{acceleration}} = \frac{\text{change in velocity}}{\text{time lapsed}} \text{ or } a = \frac{v_t - v_0}{t}$$

If the rate of change is from zero to 60 mph (100 km/h) in 30 seconds, then

$$a = \frac{(60 - 0)}{30} = 2 \text{ mph/s}$$

for a car accelerating from rest to 60 mph (100 km/h) in 30 seconds.

For those cases where the acceleration is constant, the following equations are useful:

$$v_t = v_0 + at \quad \text{and} \quad v_t{}^2 = v_0{}^2 + 2ad$$

CORE FACTS

■ Speed is the time required to travel a given distance and is expressed with a time unit and a distance unit.

■ Velocity is speed in a specified direction.

■ Acceleration is the increase of velocity; negative acceleration (or deceleration) is the decrease.

■ Acceleration and deceleration are expressed with a distance unit and the square of a time unit.

To calculate the distance traveled:

$$d = v_0t + \tfrac{1}{2}at^2 \text{ or } d = \frac{v_0 + v_t}{2}t$$

If the rate of change is from zero to 60 mph in 30 seconds, then

$$a = \frac{60 \text{ mph} - 0 \text{ mph}}{30 \text{ s}} = \frac{60 \text{ mph}}{30 \text{ s}} = 2 \text{ mph/s}$$

If the acceleration is constant, the velocity increases by 2 miles (3.2 km) per hour every second.

If the initial velocity, acceleration (or deceleration), and distance are known, the final velocity can be calculated using the equation

$$v_t^2 = v_0^2 \pm 2ad$$

Calculating motion

Any force acting on a body can cause an acceleration. In the case of gravity, a falling object, initially at rest, increases its rate of velocity by 32 ft/s (9.8 m/s) every second, which is equivalent to an acceleration of 32 fts^{-2} (9.8 ms^{-2}). If this acceleration remains constant, the distance the object falls can be calculated at each time interval during its motion using the equation

$$d = v_0t + \tfrac{1}{2}at^2$$

For example, after one second the object has fallen a distance of 16 ft (4.9 m), after two seconds it has fallen a distance of 64 ft (19.6 m), after three seconds it has fallen another 145 ft (44.1 m), and so on. These distances can be calculated by substituting the values of all the other variables into the equation above; that is the initial velocity (v_0) is equal to 0 ms^{-2}, the acceleration due to gravity (a) is equal to 9.8 ms^{-2}, and the time (t) is equal to one, two, or three seconds repectively.

The distance covered during straight-line, or linear, acceleration is fairly easy to calculate. If a basketball player drops a ball from shoulder height, gravity accelerates it. To determine the distance it travels to the floor, the fall simply has to be timed. Matters become more complicated when a change in direction occurs, too. If a basketball player drops a ball from shoulder height, 6 ft (1.8 m), the time it takes to reach the floor can be calculated:

$$d = \tfrac{1}{2}gt^2, \text{ thus } t = \sqrt{\frac{2d}{g}} = \sqrt{\frac{2 \times 6}{32}} = \sqrt{\frac{3}{8}} = 0.61 \text{ s}$$

A ball thrown straight up will rise to a height depending on its initial upward velocity, decelerating constantly at -32 fts^{-2} (-9.8 ms^{-2}), until its velocity is 0. Then it will descend as if dropped from that height.

A ball that is thrown horizontally will undergo two kinds of motion simultaneously. It will maintain its horizontal velocity while being pulled to Earth at 32 fts^{-2} (9.8 ms^{-2}). Its path will be parabolic. A ball thrown up at an angle will also execute a parabolic path as a combination of horizontal motion and upward motion, followed by downward motion. A 45-degree angle will result in maximum horizontal distance.

Suppose the basketball player tosses the ball straight up. It will rise until gravity exhausts the force of the toss, reverse direction, and come down at 32 fts^{-2} (9.8 m^{-2}). To figure out the distance, the deceleration of the rising ball and the acceleration of the falling ball both have to be measured. If the player makes a long pass down court, the ball accelerates vertically while moving at a constant velocity horizontally. It follows an arc-shaped path, or parabola, as gravity bends its trajectory to approach a straight line.

A projectile, such as a basketball, that is thrown at a 45-degree angle covers the greatest distance. One launched at an angle greater than 45 degrees and one launched at an angle less than 45 degrees cover the same horizontal distance, provided they are thrown with equal force and that the sum of the two angles is 90 degrees. Thus, a ball thrown up at 60 degrees goes the same distance as one hurled at 30 degrees.

Circular motion

A common type of motion in which speed does not change but direction does is uniform circular

CHRISTIAAN HUYGENS

In about 1583, Italian mathematician and physicist Galileo Galilei (1564–1642) is said to have used his pulse to time the swing of a lamp in Pisa Cathedral. He discovered that a pendulum—a weight swinging at one end of a tether—always swings to and fro in the same period of time regardless of the amplitude (length) of the swing. At the end of each oscillation, the pendulum has a velocity of zero, and then it accelerates to the middle due to the force of gravity.

Around 1656, Dutch mathematician and physicist Christiaan Huygens (1629–1695) applied this idea to measuring time and realized that the motion of a pendulum can be made into an exact timing mechanism if each swing is identical. His description of a clock made on this principle led almost immediately to great improvements on earlier clocks. His later studies delved deeply into the very nature of movement. He recognized that any measurement of velocity must be made in comparison to a still background, called a frame of reference, yet frames of reference are only assumed to be at rest. They cannot be proved to be standing still from every possible point of view. He thereby developed the idea of relative frames of reference, that is, motion that looks different from different points of view. A good example of this relative motion occurred to Huygens in the relation between circular motion and harmonic motion (regularly repeating motion). If a ball moves in a circle, we see this circular motion from above, but if we rotate the point of view so that we see the motion from the side, the ball seems to move back and forth, like a pendulum.

DISCOVERERS

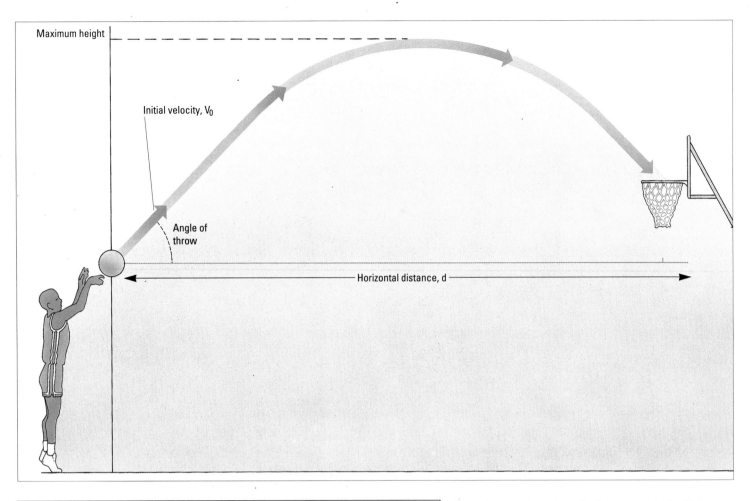

Maximum height

Initial velocity, V_0

Angle of throw

Horizontal distance, d

MEASURING MOTION

The most widely used motion-measuring device is the speedometer. In motor vehicles, it is either mechanical or electronic. In the first type, a flexible cable is fixed to the drive mechanism; as the drive turns, the cable spins a magnet, which moves a surrounding drum connected to the speedometer needle that the driver sees on the dashboard. The second type converts rotation in the transmission into electronic pulses; the pulses move a needle or are translated to digital readout on the dashboard. In both cases, engineers have taken advantage of the fact that rotational velocity, such as that of the drive shaft, is directly proportional to linear velocity: if the drive's rotational velocity doubles, the vehicle's speed doubles, too.

Airplane airspeed indicators detect changes in air pressure. A pitot tube is a U-shaped device that contains a dense liquid, such as mercury, in its elbow. One opening in the tube faces the airflow, while a second is perpendicular to it. As the airplane's speed increases, the pressure difference between the two openings moves the liquid. A pressure gauge detects the movement and, with corrections for temperature and air density, shows it as units of airspeed to the pilot.

To record a vehicle's changes in velocity, that is acceleration, a linear accelerometer comprises a weight hung from a spring and encased in a fluid-filled container. As the vehicle accelerates, the weight's inertia exerts drag on the spring, and electronic sensors convert that mechanical movement into an electronic signal. For rockets and spacecraft, the device usually indicates acceleration as multiples of Earth's gravity (symbolized by g): an acceleration of six g, for instance, means that astronauts suddenly feel six times heavier than normal.

Scientists and engineers use other instruments to measure gas flow, such as the anemometer (a shaft supporting cups arranged at right angles) and the bridled pressure plate, both of which meteorologists use to measure the force or speed of the wind. Radar and lasers can exploit the Doppler effect to measure the velocity of distant objects from the frequency of the electromagnetic waves they reflect.

A CLOSER LOOK

Three factors determine whether a basketball will drop into a hoop: the ball's initial velocity, the angle at which the ball leaves the player's hands, and the horizontal distance between the player and the hoop.

motion. A satellite in a circular motion, a race car driving around a circular track, and the atoms of a wheel rotating about a fixed axle are all engaged in uniform circular motion. The acceleration of a particle in uniform circular motion is given by

$$a = v^2 / r$$

where v is the magnitude of the particle's velocity and r is the radius of the path. Since the acceleration is directed toward the center of the circle, it is called centripetal acceleration. Producing this requires a centripetal force (see CENTRIFUGAL AND CENTRIPETAL FORCES).

R. SMITH

See also: AERODYNAMICS; CENTRIFUGAL AND CENTRIPETAL FORCES; FLUID MECHANICS; FORCES; FRICTION; MECHANICS; MOMENTUM; MOTION; RELATIVITY; TIME; TIME TRAVEL; WAVE MECHANICS.

Further reading:
Catnell, J. D. and Johnson, K. W. *Physics.* 3rd edition. New York: John Wiley & Sons, 1995.
Hazen, R. M. and Trefil, J. *Science Matters.* New York: Doubleday, 1991.
Hewitt, P. G. *Conceptual Physics.* 7th edition. New York: HarperCollins, 1993.

LAKES

A lake is a body of fresh or saline water occupying a hollow in Earth's surface

Lakes are found in all shapes and forms around the world. Some are natural and others are artificial; some contain fresh water and others salt water; some are very deep, while others are extremely shallow. Lakes certainly add to the beauty of the landscape, and they benefit humans. They provide us with fish for food, opportunities for water sports, and a source of energy for hydroelectric power production.

Some disagreement exists among hydrologists over the classification of a body of water as either a lake or an inland sea (see INLAND SEAS). In most cases, salinity and size are taken as the determining factors, with lakes usually being smaller and having a lower salinity than inland seas.

Lake locations and types

Lakes divide roughly evenly into saltwater and freshwater, with about 25,000 cubic miles (104,000 km^3)

of lake water being salty, and some 30,000 cubic miles (125,000 km^3) being fresh. Looking at a map, lakes are clearly seen to be more abundant in some areas than in others. For example, North America, and especially Canada, has an enormous number of lakes, ranging from the large to the very small, and these are found throughout the landscape. South America, in contrast, has fewer lakes, although rainfall is high over much of that continent.

The largest of all lakes is Lake Superior on the border of Canada and the United States. It has a surface area of about 32,000 sq miles (82,900 km^2). This freshwater lake forms part of the string of connected lakes known as the Great Lakes of North America. From west to east these are: Lake Superior, Lake Michigan, Lake Huron, Lake Erie, and Lake Ontario. Together, they comprise the biggest continuous area—94,700 sq miles (245,240 km^2)—and volume—6000 cubic miles (25,000 km^3)—of fresh water. However, the world's largest single freshwater lake, in terms of volume, is Lake Baikal in Russia, which contains some 5520 cubic miles (23,000 km^3), or one-fifth of all fresh water. Baikal is also the world's deepest lake, with a maximum depth of over 1 mile (1.6 km). In lakes where the inflow of fresh water is matched by the outflow, the water that accumulates in the lake will normally be fresh.

Salt lakes form when the lake is subject to extreme evaporation, a process that concentrates salts in the lake, and notably in those lakes that do not have a regular or large enough outflow and where the inflow itself may be salty; for example, Great Salt

The Great Lakes in northeastern North America is made up of five large lakes that are connected to the Atlantic Ocean by the St. Lawrence River. The lake farthest to the west is Lake Superior, which is the world's largest lake by surface area.

CORE FACTS

- Lakes form in catchment areas called drainage basins; the character of the lake depends on the geology and vegetation of the catchment area, land use, and level of pollution.
- Around half of all lakes are freshwater and around half are saltwater.
- Glaciers, ice and rock, dunes, rivers, flooding, and volcanoes can cause lakes to develop; humans can make artificial lakes—for example, a lake forms when a river is dammed to produce hydroelectric power.

CONNECTIONS

- Lakes are sometimes found in the **CRATERS** of old **VOLCANOES**.

- The effects of **DROUGHT** are seen in many lake environments, as the rate of **RIVER** water inflow declines.

Lake Baikal in Russia is the world's largest lake by volume and the world's deepest lake. It is located in a rift valley and is about 395 miles (635 km) long and 45 miles (72 km) wide.

Lake in the United States. (Great Salt Lake is referred to by some hydrologists as an inland sea.) Many of the lakes of the East African Rift Valley are not salty but contain strong solutions of sodium carbonate, which originate in nearby volcanic soils.

Structures of lakes

Lakes form in catchment areas. The drainage basin is the land from which water drains into the lake, which occupies the lowest part of the catchment area, called the lake basin. Where underlying rock is hard and insoluble, lakes tend to be clear and rather low in nutrients, and the water in them is normally fairly acidic, closely resembling the rainwater that fills them.

However, in areas of limestone rock (see LIMESTONE), which reacts with acidic water to form soluble calcium bicarbonate, the water that accumulates in lakes will be alkaline, containing a large amount of dissolved lime. This, in turn, produces a more fertile and productive lake for the plants and animals it supports. In very permeable limestone regions, permanent surface lakes tend to be rare, but underground rivers and lakes may develop when the percolating water comes up against lower, impermeable rocks.

The geology of the catchment area also has a profound influence on the type of lake that develops in it, as do other factors, such as the type of vegetation, land use, and human influences, including pollution (see POLLUTION).

Formation of lakes

Lake formation depends on a range of interrelated factors, including rainfall, geological history, gradients, and the nature of the underlying rocks and soil. Glaciers have had a major influence on the way that lakes are formed, especially in the northern temperate regions, where the Pleistocene (from 10,000 years ago to 1.6 million years ago) ice age resulted in large areas being covered with thick ice. As glaciers moved slowly along valleys, they gouged out the underlying rocks. Then, as they retreated again, the valleys remained blocked by end moraines (that is, an accumulation of material left behind), producing natural river dams that created lakes. Glacial lakes are found in areas such as the Lake District in England, and areas of Switzerland and Sweden, as well as in many parts of North America. The Okanagan and Kootenay systems in British Columbia, Canada, are good examples of this type of lake. Such lakes tend to be long and thin and to follow the contours of valleys. Another type of glacial lake is a cirque lake, which can be found in bowl-shaped hollows in mountainsides, where glaciers developed and carved out the hollows by plucking and abrasion (see GLACIERS).

Another kind of glacial lake forms by the scouring effect of ice movement over relatively flat surfaces of resistant rock. The result of this kind of ice erosion is an array of small lakes dotting the landscape, as can be seen, for example, in Finland, and also over large areas of Canada.

River systems produce lakes in a number of ways. As a lowland river flows across its floodplain, it often forms loops and meanders. When a meander in the river becomes cut off as a result of the river taking a shorter route downstream, the cutoff section of river may retain enough water to form an oxbow lake (see RIVERS).

Regular flooding of rivers can also produce lakes in a floodplain. Flooding of this type in the Amazon Basin creates a patchwork of forest, often periodically flooded, and floodplain lakes, which vary in depth depending on the season. Where a river is prevented from reaching the sea by a barrier, such as a system of wind-created dunes, a lake may develop very close to the sea. Such lakes, which may be freshwater, brackish, or saline, depending on the balance of flow between the river and sea, are called coastal lagoons.

Lagoons are sometimes tidal, and the amount of salt and fresh water varies with the sea level and the stage of the tide, as well as with changes in the flow of the river entering the lagoon. Lakes can also be formed by movements of the land. Very large lakes can result from faults in Earth's crust, as in the rift valley lakes of Africa. Africa has two great series of rift valleys: the western and eastern rifts. The west-

---

LAKE-LEVEL CHANGE AND CLIMATE CHANGE

Over time, many lakes show evidence of major changes in water levels, and geologists can use this kind of information to reconstruct the history of the lake, and also to correlate such changes with climatic fluctuations. A good example is provided by Great Salt Lake in Utah. Some 15,000 years ago, a large lake called Lake Bonneville covered 19,300 sq miles (50,000 km²) and stretched from present-day Nevada north to Idaho. Different historical lake levels are reflected in a series of parallel, raised beaches on the sides of the surrounding hills and mountains, and these changes in turn reflect changes in the climate during the period. The present Great Salt Lake is just a small part of this original giant lake.

A CLOSER LOOK

ern rift contains Lake Tanganyika, with Lake Nyasa farther south occupying another rift. The lakes of the eastern rift include those of Tanzania and Kenya (such as Lakes Manyara, Naivasha, Nakuru, Turkana, and Natron), and also a series of lakes in Ethiopia, as well as the Sea of Galilee in northeast Israel. In some cases, the land has slipped downward between faults, creating what is called a graben lake (see FAULTS). This usually has a flat bottom and steep sides. Graben lakes are often very deep. Examples include Lake Ohrid in southeastern Europe, Lake Tahoe in the United States, and Lake Baikal in Russia.

Volcanic activity also sometimes results in lakes. Perhaps the most dramatic are crater lakes, which appear when a volcanic crater fills up with water, normally after the volcano has erupted and become quiescent, leaving behind a large crater, flanked by steep, rocky sides. One of the best known is Crater Lake in Oregon. This is huge and roughly circular, about 6⅓ miles (10 km) across and about ⅓ miles (600 m) deep. Another famous crater is Ngorongoro in Tanzania. This is about 9 miles (14 km) across, and its largely flat or undulating crater base contains several lakes. In some volcanic areas, lava flows have blocked streams or rivers, causing lakes to form behind these natural dams. Examples are the lakes around the Virunga volcanoes in Africa (Rwanda, Uganda, and Zaire).

Process of change

Lakes are constantly changing, although the processes of change are often slow and hard to detect. Rivers bring sediment into the lake that grad-

LARGEST LAKES

The eight largest lakes in the world, in terms of surface area, are as follows:

Name	Area (square miles)	Area (km²)
Lake Superior	32,007	(82,900)
Lake Victoria	26,564	(68,800)
Lake Huron	23,000	(59,580)
Lake Michigan	22,401	(58,020)
Lake Tanganyika	12,703	(32,900)
Lake Baikal	12,162	(31,500)
Great Bear Lake	12,096	(31,330)
Great Slave Lake	11,031	(28,570)

ually fills it; in shallow lowland lakes this process can be relatively rapid. Silt and sediment accumulate, making the edges of the lake more shallow, eventually allowing plant growth, which further accelerates the process of infilling, known as succession. As a result of this process, aquatic lake communities change over time and become increasingly terrestrial, usually via intermediate stages of fen or swamp. The study of lake sediments can reveal much valuable information about the history of the lake, and organic remains can reveal a lot about the history of vegetation in the area.

Lakes may, over time, be made more fertile by a process called eutrophication (or accelerated productivity), often as a result of nutrient-rich runoff,

The picture below, of the Rio Villano in the Oriente rain forest of eastern Ecuador, shows an oxbow lake in the process of formation (to the left). An oxbow lake forms in the abandoned channel of a meandering river when erosion cuts off a loop in the river's course.

usually from agricultural land. In these cases, the whole natural balance of the lake is often disturbed and problems such as algal blooms may occur.

Human-made lakes

Perhaps the most prominent artificial lakes are those created by damming the flow of a river, usually to make a reservoir for supplying drinking water or as a source of hydroelectric power. Flooded valleys also enable local people to establish control over the river downstream and to prevent dangerous fluctuations in river level. Some of the largest of all artificial lakes are in Africa, and these include Lake Volta, 3275 sq miles (8515 km^2); Lake Nasser, 2399 sq miles (6216 km^2); and Lake Kariba, 2050 sq miles (5330 km^2).

Lowland reservoirs can be constructed, especially in areas with impermeable soils, by enclosing them with artificial earth or concrete banks. These spaces can then be filled up by pumping water into them or by diverting the water in rivers. The stored water is then used when it is needed.

Another distinctive kind of human-made lake is the gravel pit. These form in areas where gravel or sand deposits have been dug out. The resulting holes often fill up with water and create, in time, attractive lakes with many amenities and wildlife. The exposed gravel quickly becomes vegetated, and the water attracts many kinds of wild animals, notably aquatic insects, amphibians, fish, and birds. Many gravel workings are now protected as nature reserves.

M. WALTERS

See also: HYDROLOGY; INLAND SEAS; RIVERS; WETLANDS.

Further reading:
Geomorphology and Sedimentology of Lakes and Reservoirs. Edited by J. McManus and R. W. Duck. New York: John Wiley & Sons, 1993.

HUMAN USES OF LAKES

People have long used lakes as a source of drinking water and for irrigating crops, as well as for a source of fish for food. Many industries also rely on water extracted from lakes. With increasing population densities and intensification of farming, lakes have become more and more vulnerable to pollution, in a variety of forms, including human sewage effluent and runoff from agricultural fields.

The latter often contains high levels of nitrogen, derived from artificial fertilizers, and this can upset the ecological balance of the lake, with serious consequences for its chemistry and wildlife.

One of the many uses that humans find for lakes is the development of marinas, such as this one on Lake Powell, Utah.

SCIENCE AND SOCIETY

LANDFORMS

A landform is a distinctive topographic feature on Earth's surface

Anyone who visits Arizona's Grand Canyon cannot fail to be impressed by its magnificent scenery of cliffs, canyons, and pinnacles. These features, which have been carved from the layered rocks by the erosive powers of the Colorado River, are just a few of the many fascinating landforms that make up Earth's rich and varied landscape.

The topographic features of Earth's surface, from the smallest sand ripple to the biggest mountain range, are called landforms. Examples of landforms include mountains, valleys, caves, beaches, sand-bars, dunes, moraines, plateaus, canyons, natural arches, escarpments, and river deltas—in fact, any distinctive feature of Earth's surface. The study of the origin and evolution of landforms is called geomorphology.

Landforms can be classified as tectonic (created chiefly by large-scale Earth movements), denudational (created by the processes of weathering and erosion), or depositional (created by a buildup of sedimentary or volcanic materials). Other landforms whose origins lie outside these definitions include craters caused by meteorite impact and features built by living organisms, such as coral reefs (see the box on page 651).

Landforms are created, altered, and ultimately destroyed through the interaction of three major processes. Large-scale movements of Earth's crust, called tectonism (from the Greek word *tekton*, meaning "builder"), tend to create variations in the elevation of Earth's surface (known as relief) in the form of mountain ranges, volcanoes, rift valleys, plateaus, and basins. The forces of denudation (erosion and transportation) tend to level out these variations by eroding material from the higher areas and transporting it to lower areas. The resulting sediments are deposited in a variety of ways in drainage systems and eventually in estuaries and along the coastlines.

If allowed to run to completion, the forces of denudation would result in a flat, featureless land-

The landforms of Canyonlands National Park, Utah, have been formed over 100 million years through erosion by wind and water.

CORE FACTS

- The study of landforms is called geomorphology.
- Landforms are created, altered, and destroyed through the interacting processes of tectonism, denudation, and deposition.
- Climate influences landform evolution by determining whether fluvial, eolian, or glacial processes will dominate the shaping of the landscape.
- Rock type and structure influence landform evolution through the phenomenon of differential erosion.

CONNECTIONS

● **GLACIERS** are moving masses of land **ICE** formed from the accumulation of snow.

● **SPELEOLOGY** is the branch of **GEOLOGY** concerned with the study of **CAVES**.

Streams and rivers constantly shape the landscape by cutting down into the bedrock to produce the typical fluvial landforms of valleys and ridges.

scape. However, tectonism continually generates new relief, so that Earth's surface is a dynamic system of constantly evolving and interacting landforms.

History of ideas

Before the 18th century it was thought that Earth had been created in a state of perfection and was gradually decaying. It was not until the late 18th century that Scottish geologist James Hutton (1726–1797) proposed that Earth was immeasurably old and had

---

HUMAN LANDFORMS

A special category of biogenic landforms covers the wide range of topographic features that have been created through the activities of human beings. Human landforms include such features as highways, dams, strip mines, quarries, canals, and bomb-craters. Human activities also have a significant effect on the evolution of natural landforms. For example, the clearing of tropical rain forest leads to greatly increased erosion of the thick, tropical soil, with the development of gullied slopes and the removal of huge quantities of sediment to the sea, where it is deposited in new sandbanks and beaches.

SCIENCE AND SOCIETY

---

undergone repeated episodes of uplift and erosion over many millions of years.

In 1899, U.S. geographer and geologist William Morris Davis (1850–1934) proposed a theory of landform evolution based on a "cycle of erosion," in which the landscape passed through three stages of development, called youth, maturity, and old age. In the youthful stage, a newly uplifted area of land is eroded by streams, which carve deep, V-shaped valleys into the originally flat surface of the uplifted plateau. As the landscape reaches maturity, the valleys enlarge and intersect to create a high-relief topography of ridges and ravines. Continued erosion leads to the stage of old age, where the effects of erosion and deposition result in a level plain with sluggish, meandering rivers, minimal relief, and perhaps a few low hills marking the outcrop of more resistant rocks. The cycle could be interrupted or repeated by a further episode of uplift, which would rejuvenate the landscape.

From the 1950s on, the emergence of the plate tectonic theory (see PLATE TECTONICS) and the realization of the importance of climate in landform evolution led to the development of more sophisticated models. It was recognized that climate change and plate tectonic processes (such as continental rifting) can drastically alter a landscape. For example, global cooling, or the drift of a continent toward a polar region, can result in the formation of glaciers and the superimposition of glacial landforms on a topography that had once been shaped by rivers. Lava flows associated with continental rifting can bury an existing landscape, filling valleys, creating new uplands, and modifying the drainage pattern.

Tectonic landforms

Tectonic processes build the large-scale features of Earth's surface by the uplift or subsidence (sinking) of large areas of Earth's crust, by the faulting and thrusting of huge blocks of rock, and by the eruption of lava onto the surface.

Tectonic processes are closely related to the movements of the lithospheric plates that make up the outer shell of the planet (see EARTH, STRUCTURE OF). Volcanism, orogeny (mountain building), and thrust faulting are generally concentrated along plate margins, where one plate dips beneath another (for example, the Andes mountains), or where two continental blocks collide (for example, the Himalayas). Rift valleys and many volcanoes occur where a continent is being pulled apart (for example, the East African Rift Valley).

Orogeny is associated with crustal thickening, so the high elevation is compensated for by a "root" of less dense continental crust beneath the mountain range. As rock material is removed at the surface by erosion, the root of the mountain belt rebounds and continued erosion exposes metamorphic and igneous rocks that were once buried deep beneath Earth's surface. Eventually a state of equilibrium is achieved when the crust returns to normal thickness. Gentle uplift or subsidence of such a stable platform can lead to renewed erosion of an elevated plateau, or to the formation of a basin that may be filled by a lake or a shallow sea.

Denudational landforms

The relief that is created by tectonic processes is gradually shaped and worn down by the processes of denudation (see EROSION). Denudational landforms can be classified according to the principal agent of denudation—fluvial (rivers and streams); eolian (wind); glacial (ice); coastal (wave action, tides, and longshore currents); and groundwater. Denudational landforms can also be produced through the processes of mass movement—slumps, landslides, and rockfalls.

Climate exerts a strong influence on the evolution of denudational landforms by determining what will be the principal shaping force. For example, fluvial processes will predominate in wet climates, eolian processes in arid climates, and glacial processes in polar regions and high mountains. Rock type and structure exert control through the phenomenon of differential erosion. This occurs because some rocks are more resistant to erosion than others, and are therefore worn away at different rates.

Streams and rivers shape the landscape by cutting down into the bedrock to produce the typical fluvial landforms of valleys and ridges. The headwaters of a river system eat backward into a plateau or mountain range and transport the eroded material downstream until the slope decreases to a stage where the river can no longer carry its load of sediment. This base level can be defined by the sea or by a lake.

The pattern of drainage (that is, the shape of a river system as seen from above) may depend on the shape and structure of the underlying rocks. On a flat surface with no structural control, the pattern is dendritic (that is, like the branches of a tree). A radial pattern (like the spokes of a wheel) develops on isolated peaks (for example, on volcanoes), and a trellis pattern is produced by the differential erosion of tilted rock strata.

Eolian processes are characteristic of arid, desert regions, where the wind winnows the sediments produced by weathering and transported by flash floods and seasonal streams. Fine silt may be blown

BIOGENIC LANDFORMS

Limestone reefs are among the biggest structures ever built by living organisms. The Great Barrier Reef stretches for over 1250 miles (2000 km) along the northeast coast of Australia and covers an area of 80,000 sq miles (207,000 km²). The reef has been growing for 25 million years, building slowly upward as the underlying continental shelf subsides slowly beneath the Pacific Ocean.

The reef's foundation is composed of limestone, which has been deposited by countless millions of tiny animals called coral polyps. The polyps, which are like tiny sea anemones, live in colonies that build coral "heads" by secreting calcium carbonate. The coral heads can be several feet across and are shaped like domes, sheets, or branches. Coral reefs provide food and shelter for many other species, such as calcareous algae, sponges, and bryozoans, whose remains, together with fragments of dead coral, contribute the cement that binds the reef together, making it resistant to the pounding of ocean surf.

The Great Barrier Reef in Australia supports a variety of different corals that grow at depths of 10 to 22 ft (3 to 7 m).

A CLOSER LOOK

The rise and fall of the sea and the power of waves are constantly reshaping the thousands of miles of Earth's coastlines. Cliff faces retreat as the wave energy releases its erosive power on the shoreline.

many hundreds of miles and deposited in another region as loess, or blown out to sea. Boulders that are too heavy to be moved may be sand-blasted by wind-carried sediment and form a desert pavement.

Glacial landforms are produced by the grinding of glaciers and ice sheets (see GLACIERS). These landforms can persist for thousands of years after the ice has disappeared. Typical glacial landforms include U-shaped valleys, cirques (circular hollows on mountainsides), hanging valleys, arêtes (narrow, jagged ridges), and striated rock surfaces.

Coastal landforms include erosional features such as sea cliffs, stacks (isolated rocky islands), and arches and depositional features such as beaches, sandbars, and spits (see COASTS). Groundwater plays an important role in limestone regions, where it dissolves the bedrock to produce caves and caverns and associated karst (solution) scenery (see CAVES; LIMESTONE).

Depositional landforms

Depositional landforms include features such as the coastal features mentioned earlier (see COASTS); moraines (heaps of clay, boulders, and gravel dumped by retreating glaciers); drumlins (stream-lined mounds of boulder clay shaped by flowing ice); and eskers (sinuous ridges of sand and gravel deposited by streams flowing beneath a glacier; see GLACIERS). Other depositional landforms result from river sytems and include floodplain deposits, alluvial fans, and deltas (see DELTAS; RIVERS). Volcanic activity can also produce thick accumulations of vol-

canic lava, such as the Columbia River basalts in the northwestern United States or cones of volcanic ash (see VOLCANOES).

Other landforms

Impact craters are circular depressions, surrounded by a raised rim, that have been created by a meteorite, comet, or asteroid slamming into the surface of Earth. Although Earth was bombarded by many meteorites in the distant past, few craters survive because the processes of erosion and orogeny have obliterated them.

Biogenic landforms are created by the actions of living organisms. Most of these are small-scale and rather short-lived, though some, such as termite mounds and beaver dams, may survive for hundreds of years. The largest and most enduring biogenic landforms are coral reefs.

N. WILSON

See also: CANYONS; CAVES; CLIFFS; COASTS; CRATERS; DELTAS; DESERTS; EROSION; GLACIERS; GROUNDWATER; MOUNTAINS; RIVERS; VALLEYS.

Further reading:

Blume, H. *Color Atlas of the Surface Forms of the Earth.* Cambridge, Massachusetts: Harvard University Press, 1994.
Summerfield, M. A. *Global Geomorphology: An Introduction to the Study of Landforms.* New York: John Wiley & Sons, 1991.

LANTHANIDE ELEMENTS

Lanthanide elements are the metallic elements from lanthanum to lutetium

The lanthanide elements are also called the rare earth metals. In the 18th century, when the only known way of extracting metals from their ores was by smelting (see SMELTING), a number of minerals were known that could not be smelted. These included lime, magnesia, and silica—which French chemist Antoine Lavoisier (1743–1794) incorrectly decided were elements—and were generally grouped together as "earths."

In 1794, a Finnish chemist, Johan Gadolin (1760–1852), discovered a new "earth" in a rock found at a Swedish village called Ytterby; he named it yttria. Some time later, German chemist Martin Klaproth (1743–1817) decided that yttria consisted of two minerals, the second of which he named ceria; and subsequently Swedish chemist Carl Mosander (1797–1858) discovered that there was a whole series of different "rare earths." The "rare earths" themselves were eventually identified as the oxides of the metals that make up the lanthanide elements.

In 1911, British physicist Charles Barkla (1877–1944) discovered that when X rays were scattered by a metal, the emission spectra were characteristic for each particular metal. Using this technique, another British physicist, Henry Moseley (1887–1915), was able to determine the atomic numbers of 14 rare earth metals: 57-lanthanum (from the Greek meaning "hidden"); 58-cerium; 59-praseodymium (from the Greek meaning "green twin"); 60-neodymium ("new twin"); 62-samarium (from *samarskite*, the mineral in which it was found); 63-europium; 64-gadolinium (in honor of Johan Gadolin); 65-terbium (from *Ytterby*); 66-dysprosium (from the Greek meaning "hard to obtain"); 67-holmium (from *Stockholm*); 68-erbium (from *Ytterby*); 69-thulium (from an old name for the northern most country); 70-ytterbium; and 71-lutetium (from an old name for Paris). The missing element, number 61, was discovered in 1945 among the products of uranium fission (see FISSION) and named promethium.

Electronic configuration

The calculation of the atomic numbers of these elements caused a problem, because they could not be

Monazite, a rare earth phosphate, contains varying proportions of cerium, lanthanum, neodymium, and thorium phosphates.

made to fit into the periodic table. Russian chemist Dmitry Mendeleyev (1834–1907) had been able to draw up this table by grouping elements with similar chemical properties one under the other. Lanthanum resembles element 39, yttrium, which comes above it in the table. But cerium does not resemble element 40, zirconium. In fact, all 14 (later 15) rare earth elements were found to have very similar properties, and it is element 72, hafnium, that resembles zirconium.

Hafnium, in fact, had already been discovered to belong to another group of metals that could not be fitted comfortably into the periodic table. These were the "common" metals titanium (22) through zinc (30), zirconium (40) through cadmium (48), and hafnium through mercury (80). These are now known as the transition metals (see TRANSITION ELEMENTS).

CORE FACTS

■ The lanthanides are the metallic elements from lanthanum to lutetium.

■ The normal source of lanthanide elements is the mineral monazite.

■ Lanthanum, the first lanthanide element, is used in the manufacture of steels and magnesium and aluminum alloys.

■ Lanthanide metals are shiny and will oxidize at different rates when exposed to oxygen or air.

CONNECTIONS

● The study of the properties of lanthanide **ELEMENTS** has been important in an understanding of the properties of **METALS**.

Following early work on the structure of the atom by Danish physicist Niels Bohr (1885–1962), it was believed that electrons surrounded the nucleus in a series of concentric shells. The innermost shell was named the K-shell; then came the L-shell, the M-shell, and so on. In 1916, U.S. chemists Gilbert Lewis (1875–1946) and Irving Langmuir (1881–1957) of the General Electric Company explained the way in which the elements were built up: the K-shell could contain one or two electrons, and successive shells could contain from one to eight. This scheme worked very well for the lighter elements up to argon and for the "typical" elements of the eight groups of the periodic table, but it did not explain the existence, and the properties, of the transition and lanthanide metals.

In 1921, C. R. Bury (1890–1968) suggested that, although eight electrons is the maximum number of electrons in the outermost shell, additional electrons could be accepted into shells inside it after the outer shell is filled. This meant that the M-shell could contain up to 18 electrons, the N-shell up to 32, and so on. This theory, which was further developed by Bohr, at once explained the properties of the transition and lanthanide elements.

Lanthanum can be visualized as having two electrons in its K-shell, 8 in its L-shell, 18 in its M-shell, 18 in its N-shell, 9 in its O-shell, and two in its P-shell. The lanthanide elements then add electrons in the N-shell, until lutetium, which has 32. These inner electrons scarcely affect the chemical properties of the elements, although their physical properties differ.

Far from rare

The rare earths are not as rare as they were once thought to be. Earth's crust contains three times as much cerium as lead, for example; and thulium, the rarest, is more plentiful than silver, gold, or platinum. The lanthanides are also found in meteorites, the Moon, Sun, and stars. One of the major sources is the mineral monazite, which contains about 50 percent by weight of the rare earth elements.

Because of their very similar chemical properties, the elements were difficult to isolate—in fact, there was very little need to do so until it became necessary to separate the products of uranium fission in the 1940s. It was then that American chemist Harold Spedding discovered that ion-exchange—a sophisticated form of water-softening—would separate the lanthanides, and that their salts could be washed from the ion-exchange column in the reverse order of their atomic numbers.

Once the lanthanides could be separated in this way, it was possible to prepare the individual metals. They are all bright and silvery but behave in different ways when exposed to air. Europium, for example, will convert entirely, within a few days, to the oxide; and lanthanum, cerium, praseodymium, and neodymium will also corrode rapidly. Yttrium, gadolinium, and lutetium, however, will remain uncorroded for years.

B. INNES

See also: ATOMS; ELEMENTS; METALS; PERIODIC TABLE.

Further reading:

Barrett, J. R. *Understanding Inorganic Chemistry*. New York: Ellis Horwood, 1991.
Cotton, F. A. *Basic Inorganic Chemistry*. 3rd edition. New York: John Wiley & Sons, 1995.

USES OF LANTHANIDES

Very few uses have so far been found for the individual lanthanide metals and their compounds, but unpurified mixtures of their oxides are important in industry. Among the first applications was by Austrian chemist Carl Auer von Welsbach (1858–1929). He invented the incandescent mantle, a fabric network impregnated with cerium and thorium nitrates, which glowed brightly in a gas flame and is still used in camping lanterns; and misch metal, a mixture of lanthanides that, combined with iron, is the material of flints for cigarette and gas lighters (see below).

Millions of tons of lanthanide minerals are used annually in the production of catalysts for the cracking of crude petroleum (see PETROCHEMICALS). Lanthanides are also important in the glass industry and in the metal industry, where they have long been added to alloys and other metals to remove impurities. A recent development is the use of a mixture of yttrium and europium oxides, sometimes combined with gadolinium oxide, to make a phosphor for television tubes that provides the red color.

SCIENCE AND SOCIETY

LASERS AND MASERS

Lasers and masers are devices that produce coherent beams of electromagnetic radiation

These sparks are produced by a laser cutting through the thin steel plate.

Lasers are devices that produce monochromatic and coherent light—that is, light of the same wavelength, with waves that have all their crests and troughs in step. Such waves are said to be in phase. A maser is a similar device, producing radiation in the microwave range.

The word *laser* stands for "light amplification by stimulated emission of radiation," while *maser* stands for "microwave amplification by stimulated emission of radiation." Masers were created first; lasers were originally called optical masers. As the name suggests, masers were designed to be amplifiers. A small amount of radiation fed into one of these devices can release powerful radiation at the same wavelength. And just as an electronic circuit that amplifies voltage can be changed into an oscillator that generates a signal, a laser or maser can be used to generate light or microwaves. Today, the names *laser* and *maser* are used for oscillators as well as amplifiers.

Radiation from atoms

Lasers and masers are really devices that convert other forms of energy into electromagnetic radiation, using atoms as the converters.

Light is composed of tiny packets of energy called photons (see LIGHT). These can behave as if they are waves as well as particles, so every photon has a wavelength. This is related to the amount of energy the photon possesses: a photon of a certain energy always has the same wavelength. The higher the energy, the shorter the wavelength.

CORE FACTS

- Lasers emit monochromatic, coherent light.
- Masers produce radiation in the microwave range.
- Both lasers and masers rely on the process of stimulated emission of radiation.
- There are many different types of lasers: gas lasers, solid-state lasers, chemical lasers, dye lasers, excimer lasers, free electron lasers, and semiconductor lasers.
- Lasers are used in welding, for cutting metal, and in communications. Masers are used as amplifiers.

CONNECTIONS

- Gas lasers use a current of **ELECTRICITY** to excite the **ATOMS** in **GASES**.

- The development of **HOLOGRAPHY** was only possible through the development of lasers.

STAR WARS

One of the most publicized proposals for the use of lasers has been as space-based weapons that shoot down ballistic missiles before they can reach their targets. This is officially called the Strategic Defense Initiative but is more popularly known as Star Wars.

One version of this plan called for huge chemical lasers, each about the size of a railroad car, in orbit around Earth. The powerful laser beam would be aimed by a movable mirror at targets detected by radar. Another version would use satellites containing hundreds of crystal rods arranged around a hydrogen bomb. When missiles were detected, computers would aim the rods at them, then the bomb would be exploded; in the instant before the rods were vaporized, they would convert some of the energy of the bomb to extremely powerful laser beams.

Critics pointed out that laser beams would lose much of their energy as they passed through the atmosphere, and they said that it would be almost impossible to aim them accurately at hundreds or thousands of missiles in the short time the missiles would spend above the atmosphere en route to their targets. They also said that missiles could easily be protected by polishing their surfaces to make them reflect laser beams.

In an atom, the negatively charged electrons are arranged around the positively charged nucleus in shells. To move from an inner to an outer shell, an electron must acquire some extra energy. You can think of this as the energy needed to move the electron away from the electrostatic attraction of the nucleus, just as it takes energy to lift a satellite into a higher orbit against the gravitational attraction of Earth.

When an electron falls back to a lower shell, it gives up energy. According to the theory of quantum mechanics (see QUANTUM THEORY), the energy needed to move from one shell to another in any atom is always the same. This energy must be added

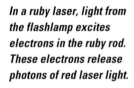

In a ruby laser, light from the flashlamp excites electrons in the ruby rod. These electrons release photons of red laser light.

or released in fixed amounts called quanta. These electron jumps are called changes of energy level.

One way an atom can acquire the quantum of energy needed to move an electron to a higher energy level is by absorbing the energy of a photon. An atom in this state is said to be excited. Conversely, when an electron falls to a lower energy level, the atom can dispose of the excess energy by releasing a photon. Since the amount of energy in these jumps is fixed, an electron can only jump up a level when the atom acquires a photon that has the same energy as is needed for the jump. And when an electron falls back, it releases a photon with the same amount of energy it took to jump up.

If a second photon arrives at the moment the atom is excited, the atom will release a photon that is in phase with it. The emitted radiation increases the energy of the passing wave. This process can go on, stimulating the emission of other photons.

In most materials there are always a few atoms in an excited state, but if photons pass through such a material, most will be absorbed by unexcited atoms. In order for a laser to act, the population of excited atoms that can emit photons must be larger than that of the unexcited atoms that can absorb them. This condition is called a population inversion. To create a population inversion, energy must be added to the system. This is called pumping the laser.

How a laser is constructed

A laser consists of a material capable of emitting photons, an energy source, and a method for trapping and directing the emitted radiation. The first lasers were made of rods of ruby. A mirror is placed at one end of the rod, and a half-silvered mirror (which lets

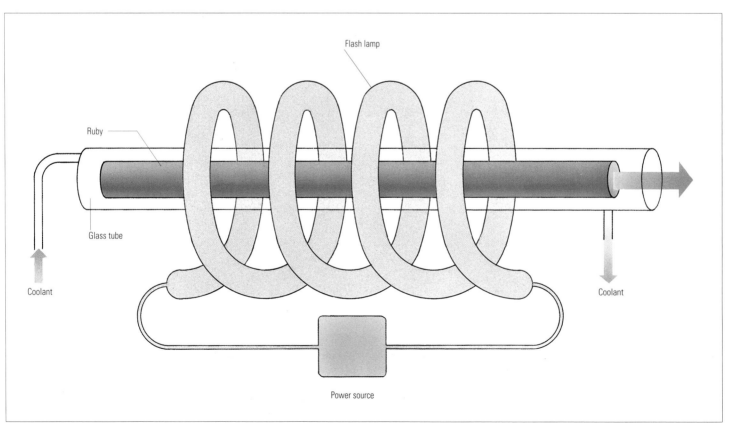

Flash lamp

Ruby

Glass tube

Coolant

Coolant

Power source

some light pass through) is placed at the other end. The rod is surrounded by a bright light source—often a group of flashtubes—as an energy source.

When the flashtubes are fired, some of their light energy is absorbed by atoms in the ruby crystal, causing their electrons to jump to higher energy states. Electrons in the excited atoms naturally begin to fall back to their normal shells, releasing photons of red light. At first, these photons are released in all directions. But those that are released along the axis of the rod are reflected back by the mirrors at the ends. These photons strike other excited atoms and stimulate these to release more photons, which travel in the same direction along the axis. These photons, in turn, are reflected back, causing still more to be released; a chain reaction is created. Almost instantly after the flashtubes are fired, a pulse of red light escapes through the half-silvered end of the laser rod.

This was how the first laser was made, but today there are several different types of lasers, which can be classified by the type of lasing material.

Types of lasers

Gas lasers use an electric current passing through a gas to excite the atoms. A current passing through a neon sign excites atoms of gas to emit light in all directions; a gas discharge laser uses the same excitation but forces the atoms to emit light along an axis. The most common gas laser uses a mixture of helium and neon and emits red light. Energy is supplied by passing a direct current through the gas; this excites the helium atoms, which in turn pass energy to the neon atoms, causing them to emit red light. The helium-neon laser is widely used in supermarket scanners and in classroom pointers. Another gas laser uses carbon dioxide and is capable of creating very high powered beams (up to at least 30 kilowatts) in the infrared range, which can be used for welding and cutting.

Solid-state lasers use crystalline materials such as ruby. A more popular form today uses a crystal of yttrium aluminum garnet, or YAG, in which the active atoms are neodymium. YAG lasers can produce fairly high power infrared beams for welding and cutting. A solid-state laser is very rugged, but the solid material heats up, since it absorbs energy while being pumped, so it must be cooled down.

Chemical lasers use the energy released by chemical reactions to power the laser. Since the lasing material is used up by the reaction, chemical lasers are either "one-shot" devices or must be reloaded with the chemical reactants.

Dye lasers are made of organic dyes, which are usually suspended in water solution. The complex dye molecules can be made to emit photons in a wide range of wavelengths, and varying the composition of the solution allows these lasers to be "tuned" over an even wider range, making light of many colors.

Excimer lasers use mixtures of chemically active gases, such as fluorine, with inert gases, such as argon. These gases can bond with each other in a special arrangement called a dimer. *Excimer* is short for "excited dimer." These dimers are capable of emit-

PHOTOCHEMISTRY

When the electrons in an atom or molecule are excited to higher energy levels, changes can occur in chemical bonding and reactivity that would not be possible without the light excitation. The study of reactions that take place in the presence of light is called photochemistry.

Laser light makes possible photochemical reactions that would not take place under ordinary light. Under such high energies, two or more electrons in a single atom can be raised to higher energy levels at the same time.

One area in which laser photochemistry seems especially promising is in the separation of different isotopes of the same element (see ISOTOPES). Ordinarily, different isotopes of a single element react in exactly the same way chemically, so they cannot be separated by chemical reactions. However, light from a laser has such a narrow range of frequencies that it can be made to excite atoms of one isotope, while leaving another unaffected. (Ordinary light sources produce a much wider range of frequencies so cannot be used selectively to excite one isotope without affecting others.) This means that just one isotope can be made to enter into a chemical reaction. This approach is being tested as a way to separate uranium-235 from uranium-238. Ordinary uranium contains many U^{238} atoms and just a few U^{235} atoms; it must be enriched with more U^{235} to work in a nuclear reactor (see FISSION).

ting laser light in the ultraviolet range, which does not cause heating, so they are valuable in delicate surgery. The short wavelength of ultraviolet light also allows it to be focused onto very tiny spots.

The free electron laser relies on the fact that when a moving charged particle is forced to change direction, energy is released in the form of a photon. In this laser, an electron beam is passed through a rapidly changing magnetic field that causes the electrons to "wiggle," releasing photons whose energy is controlled by the frequency of the changes in the magnetic field. This laser can be tuned to almost any wavelength.

This woman is being treated by lasers for cancer of the throat. Red argon laser light, traveling through four optical fiber waveguides, activates a drug that is toxic to cancerous cells.

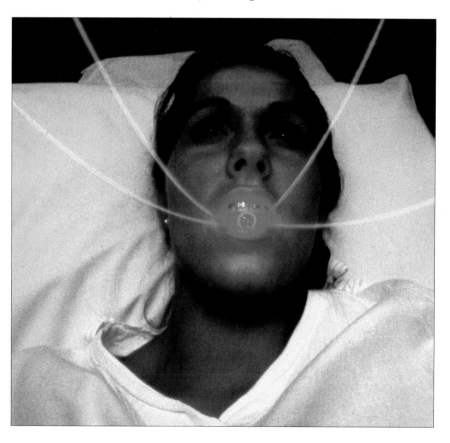

Semiconductor lasers convert an electric current into laser light. These are low-powered lasers and are used in devices such as computer printers and light sources for optical fiber circuits.

Lasers can emit continuous or pulsed radiation. Pulses are made by putting a shutter between the lasing material and the end mirrors. This can be mechanical or an opaque solid or liquid that turns clear when an electric signal is sent to it.

How masers are constructed

The first masers used excited gases. In one design, a gas is driven through a small opening into a space where electric fields separate excited atoms and unexcited ones, creating a population inversion (a system of atoms that has more excited atoms than unexcited atoms). The excited atoms then pass into a space whose dimensions are an exact multiple of the microwave wavelength to be amplified. The dimensions of the space create a resonance that causes photons of the desired wavelength to bounce back and forth, stimulating the emission of more photons.

Newer masers use crystals such as ruby. The atoms in crystals are in the form of ions; that is, they are electrically charged. It is the attraction between these charged atoms that holds the crystal together. Magnetic fields can be applied to the crystal to change the energy levels possible in ions, and in this way the maser can be tuned over a range of frequencies.

Applications

Because light emerges from a laser in a tight, coherent beam rather than radiating in all directions, its intensity does not diminish with distance. This makes lasers ideal for communication. A laser beam can be modulated like a radio carrier wave to carry sound or picture information, or it can be broken into pulses to carry information in digital form. Usually, laser light is used to send data through optical fibers.

The laser can also deliver lots of energy in a very narrow beam, supplying heat for welding and cutting in a much finer line than is possible with any kind of torch or electric arc system. Scientists believe they someday may be able to use lasers to heat hydrogen molecules to a temperature where they will fuse into helium, forming an almost unlimited source of energy (see HYDROGEN).

Lasers are replacing scalpels for some types of surgery. At the same time a laser beam cuts flesh, it can also cauterize (burn) the wound, sealing it to prevent bleeding. Laser beams can also be focused through the transparent lens of the eye to perform delicate surgery on the retina.

At the other end of the energy scale, lasers play an increasingly important part in bar-code readers, CD players, and computer printers. Without lasers, the current state of development of holography (see HOLOGRAPHY) would have been impossible.

Masers are more often used as amplifiers. Electronic circuits always introduce some "noise," because the thermal motion of the atoms in conductors and other circuit elements creates signals that mix with, and sometimes overpower, the signals to be amplified. Because a maser amplifies only a single frequency, it is a very low noise amplifier. Maser amplifiers are used, for example, to amplify the very weak signals received by radio telescopes or in long-distance microwave communications.

W. STEELE

See also: ATOMS; ELECTRONS AND POSITRONS; LIGHT; QUANTUM THEORY.

Further reading:
Agrawal, G. P. *Semiconductor Lasers*. New York: Van Nostrand Reinhold, 1993.
Elitzer, M. *Astronomical Masers*. Boston: Kluwer Academic Publishers, 1992.

ATOMIC CLOCKS

Every clock works by counting some sort of ticks. Early clocks used a pendulum adjusted to swing back and forth once a second. Mechanical clocks and watches use a weighted wheel that spins back and forth against a spring. Digital clocks and watches, whether they use a digital (numerical) or analog (traditional clock face) display, use a small vibrating crystal to generate a radio-frequency signal, which is then divided repeatedly by an electronic circuit into pulses one second long. None of these devices can be more accurate than the system that supplies the "ticks." A typical digital clock may gain or lose a few seconds a month, because the crystal that generates the signal is affected by temperature and other changes.

The first atomic clock was built in 1956 and is accurate to about one second in 300,000 years. It uses a maser based on cesium atoms to produce a signal with a frequency of 9192.63177 MHz, which can be divided down to human counting speeds. Since the photons emitted by cesium atoms are always of exactly the same frequency, the "ticks" it produces never vary in length. The cesium atom is now defined as the international standard of frequency, and the U.S. Naval Observatory uses several cesium clocks as the basis for its radio time signals.

Other atomic clocks use masers based on rubidium and hydrogen atoms. Masers like those in these clocks are also used as frequency standards for tuning radio transmitters and other electronic instruments.

This cesium atomic clock at the German National Standards Laboratory, near Hanover, is the world's most accurate clock, losing about one second every million years.

LATENT HEAT

Latent heat is the heat gained or lost by a substance, without a temperature change, during a change of state

When you climb out of a swimming pool, even on a hot day, you feel your body getting cool. This is because water is being evaporated (see EVAPORATION) by heat from your body. This heat energy allows the water molecules on your skin to break away from each other. As the water molecules move from the liquid to the gaseous state (evaporate), they take energy with them in a form of stored or latent heat, and your skin is cooled down. When the molecules condense back to a liquid somewhere else, they will release this heat again.

Latent heat is the heat gained or given out by a substance during a change of state, such as when a solid changes into a liquid. If the pressure remains constant, the temperature of the substance will not change with the gain or loss of latent heat, because the heat is stored or released in the process of change of state, or phase change. The amount of heat energy stored or released as latent heat is usually measured in joules or kilojoules.

Scottish chemist and physicist Joseph Black (1728–1799) established the concept of latent heat in 1761. In his experiments, Black timed how long it took to heat a given mass of water from 33°F (0.56°C; just above freezing point) to 34°F (1.1°C). He then compared the time it took to melt the same amount of ice. He found it took longer to melt the ice because heat was absorbed as latent heat during the phase change. During this phase change, he saw that the temperature of the substance did not change.

Temperature experimentation

Black also investigated the latent heat associated with changing water to a vapor. He heated vessels containing measured amounts of water and timed how long it took the water temperature to change from 50°F (10°C) to the boiling point. He compared this with the time it took the water to boil away. From his experiments, Black calculated that the amount of heat needed to boil the water away was the same as that needed to raise the temperature of the same amount of water by over 1000°F (538°C). To put it another way, over six times as much heat is needed to vaporize a given quantity of water than is needed to raise it from just above freezing point to boiling point.

This athlete uses the latent heat of vaporization to cool down. When liquid water touches him, it absorbs heat energy from his skin and then vaporizes.

The heat of vaporization is the latent heat associated with changing a liquid or a solid to the gaseous state, or condensing a substance from the gaseous state to the liquid, without a change in temperature. The heat of fusion is the latent heat associated with melting a solid or freezing a liquid.

To change the state of a substance to a liquid or a gas, the attraction between the molecules that keeps them a solid or a liquid must be overcome. Latent heat is the energy that overcomes these forces of attraction, and it is proportional to the strength of these

---

CORE FACTS

- A change of state—from solid to liquid or liquid to gas—requires energy, in the form of heat, to increase the separation of molecules. This heat is called latent heat.
- While a substance is going through a change of state (or phase), its temperature does not change.
- The latent heat absorbed during melting or vaporization is emitted again when the change of state goes in the reverse direction.

CONNECTIONS

● When **CARBON DIOXIDE** changes directly from a solid to a gas, it absorbs latent **HEAT**.

● **WATER** has a high **HEAT** of vaporization due to **CHEMICAL BONDS** between its molecules.

In the period from point B to C on this cooling curve for naphthalene (right), the liquid state releases latent heat but does not change in temperature. The phase diagram for carbon dioxide (below) shows the temperature and pressure at which the gaseous, liquid, and solid states are stable.

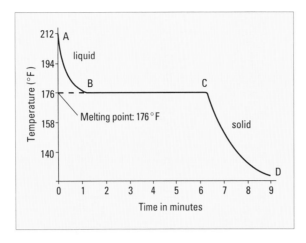

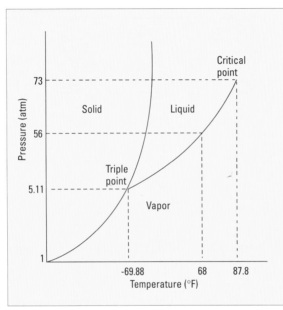

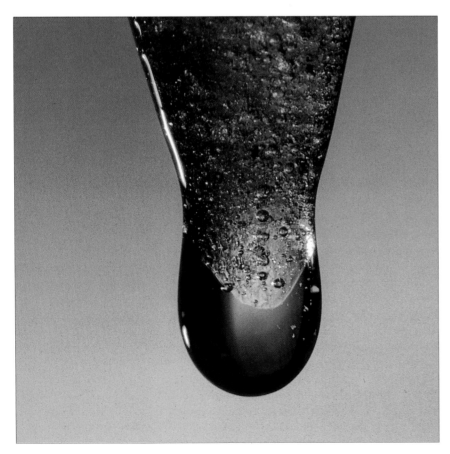

forces. To isolate the effect of molecular weight, scientists calculate the molar heat of a change of state—that is, the heat absorbed in melting or vaporizing one mole (6.02×10^{23} molecules) of the substance.

Vapor rules

Trouton's rule states that when liquids are vaporized, the ratio of the molar heat of vaporization to its boiling point is a constant. This rule is based on the assumption that the entropy of vaporization is constant, which is explained by the fact that there is an equal increase in separation and randomness of molecules when vaporization occurs. This means that the entropy of vaporization is a constant (see ENTROPY).

A notable exception to Trouton's rule is water. The latent heat of vaporization of water is abnormally high because of the extra attractive forces between its molecules. These forces are due to very strong hydrogen bonding (see CHEMICAL BONDS). The large amount of latent energy given up by condensing steam is exploited in industry: it is used for power generation and for heating plant equipment such as dryers and distillation columns.

Some substances, such as carbon dioxide, do not normally have a liquid phase but pass directly from solid to gas. This is called sublimation. The latent heat associated with phase changes for a substance is constant. This means that the same amount of latent heat is stored or released whether the substance passes from the solid to the liquid state or sublimes directly to the gaseous state. On a cold, dry winter day, even snow can go from the solid to gaseous state without going through a liquid state.

Temperature and pressure affect the characteristics of a substance. They determine whether it will be in the solid, liquid, or gaseous state. The conditions under which vaporization or condensation takes place are determined by vapor pressure (see GASES). The relationship between the temperature and vapor pressure of a liquid can be shown on a phase diagram (see the diagram at left, center). A phase diagram shows under what conditions a particular substance will remain as a liquid, solid, or gas. The plot of temperature versus pressure shows the conditions at which the solid, liquid, or gaseous phase of the substance is in a thermodynamically stable form.

P. WEIS-TAYLOR

See also: EVAPORATION; GASES; LIQUIDS; MATTER; MELTING AND BOILING POINTS; SOLIDS.

Further reading:

Gardner, R. and Kemer, E. *Science Projects About Temperature and Heat*. Hillside: Enslow, 1994.
Hazen, R. M. and Trefil, J. *Science Matters*. New York: Doubleday, 1991.
Lock, G. S. H. *Latent Heat Transfer: An Introduction to Fundamentals*. New York: Oxford University Press, 1994.

When the air warms to above freezing point, this icicle will extract heat energy from it and the solid ice will melt.

LEAD

Lead is a blue-gray metal in group 14 of the periodic table

The metal lead has been known since early times. The ancient Egyptians used it in a variety of ways, and the Romans made water pipes from the metal. In fact, it is from the Latin word *plumbum*, meaning pipe, that we get the word *plumbing* and the chemical symbol for lead, Pb.

Occurrence and properties of lead

Lead is seldom found as a free element in nature, and even its mineral compounds are relatively rare. Scientists have calculated that lead forms about 15 parts per million of Earth's crust. Fortunately, the most important ore—galena, the sulfide PbS—is found concentrated in various deposits in places throughout the world. Because the sulfide is easily oxidized into sulfur dioxide, and because lead has a relatively low melting point of 621°F (327°C), ancient people could obtain the metal just by heating galena on an open hearth:

$$PbS + O_2 \rightarrow Pb + SO_2$$

Lead is a blue-gray metal. A newly cut surface is shiny, but in air it will rapidly become coated with a dull thin film of oxide. In its usual form, lead is soft and malleable. However, there is also a close-packed crystalline form of lead. If a zinc rod is hung in a weak solution of lead acetate, the zinc replaces the lead in solution, and a lead tree—a mass of crystals—forms on the rod.

Lead is one of the group 14 elements. This group also contains carbon, silicon, germanium, and tin. Lead has oxidation states of +2 and +4, but most compounds are of the Pb(II) state. Because of its protective oxide film, lead is not easily attacked by concentrated acids, which has made it very useful for lining storage tanks. However, it is gradually attacked by water that contains dissolved oxygen, and for this reason the metal is no longer used in domestic piping.

There are three oxides of lead. The most stable is litharge (PbO), a yellow or reddish yellow solid. Slow heating of litharge in air gives red lead, or minium (Pb_3O_4), a bright red crystalline powder that is used as an antirust pigment in paints. Oxidation of red lead by nitric acid gives lead dioxide (PbO_2). This brown powder is a powerful oxidizing agent, and many inflammable substances will

CORE FACTS

- Lead is a soft, malleable blue-gray metal.
- Lead is easily oxidized. There are three oxides of lead: litharge (PbO); minium (Pb_3O_4); and lead dioxide (PbO_2).
- Lead is heavy, so it is used in shot; it is malleable, so it is used in solder.
- Lead is extracted by the blast furnace smelting of the mineral galena.

The mineral galena is one of the most important ores of lead. It contains cubic lead-gray crystals.

explode or burn violently when they come into contact with it. It is used in match heads. It is also used in lead-acid batteries; one plate is made of lead dioxide and the other of lead (see BATTERIES).

Most Pb(II) salts are insoluble, white, or colorless. Lead carbonate, or white lead, is found as the mineral cerussite ($PbCO_3$) and generally occurs as the basic compound ($PbCO_3)_2.Pb(OH)_2$. Lead chromate or chrome yellow ($PbCrO_4$) is used as a paint pigment. It was used for more than a century as a paint pigment because of its whiteness and ability to coat. But it had a major disadvantage: lead poisoning. The symptoms of lead poisoning include constipation, vomiting, headaches, tiredness, and irritability. Children who eat paint flakes are particularly at risk. In the 1960s, this caused an outbreak of a mysterious brain disease. As a result, lead compounds were banned from paints used inside buildings.

Among the few soluble lead compounds, lead acetate, $Pb(C_2H_3O_2)_2$, is used in dyeing as a mordant. Mordants help the dye stick to the fibers of the cloth (see DYES). Lead azide, $Pb(N_3)_2$, is highly explosive and is used in detonators (see EXPLOSIVES).

Uses of lead

Lead is a heavy metal, with a density of 11.35 g/cm³ at 68°F (20°C). It is used in applications where weight with minimum bulk is required. Shot is made by dripping molten lead from a height into a tank

Symbol: Pb
Atomic number: 82
Atomic mass: 207.2
Isotopes:
 204 (1.4 percent),
 206 (24.1 percent),
 207 (22.1 percent),
 208 (52.4 percent)
Electronic shell structure:
 $[Xe]4f^{14}5d^{10}6s^26p^2$

CONNECTIONS

- Lead is found in the same group as **CARBON, TIN,** and **SILICON** in the **PERIODIC TABLE**.

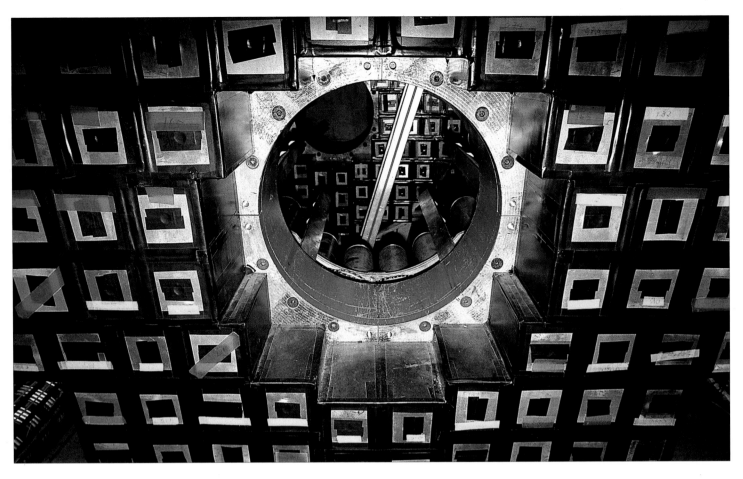

These lead glass blocks form part of one of the end plates in a particle detector.

filled with water or oil. As they cool in falling, the drops of metal become spherical. Lead remains an important part of solder because of its malleability. Until the development of computerized typesetting, lead was also important in casting type. Nowadays,

two-thirds of the lead produced in the United States is used in electric storage batteries. Until recently, the compound tetraethyl lead was added to gasoline to reduce knocking, the sound that occurs with explosive spark ignition (see HYDROCARBONS), and to lubricate upper parts of the engine, such as the valves. Leaded gasoline has been phased out in the United States because of fears about lead poisoning.

Extraction from galena

The primitive method of extracting lead from galena was to heat it in air on the bed of a furnace, which formed sulfur dioxide. Lime was then added to form a slag with any other minerals present, which melted on top of the lead and could be skimmed off.

Because of the labor-intensiveness of this process, it was replaced by blast furnace smelting (see SMELTING). To remove other minerals, the ore is first put through a flotation process. It is crushed, then mixed with water and oil; the sulfide particles stick to the oil and float to the top, while the unwanted material sinks to the bottom. The concentrate is dried, mixed with coke and lime, and reduced in a blast furnace.

Lead ores contain important quantities of other elements: much of the world's bismuth, arsenic, antimony, and silver is obtained from the refining of lead.

B. INNES/E. KELLY

RADIOACTIVE LEAD

The alchemists of the Middle Ages believed that lead was the oldest of metals. They thought that the other metals had developed from it, and that they themselves could discover how to "breed" gold. Nobody now considers this possible, but early in the 20th century, scientists made a discovery that seemed just as unlikely: a whole handful of elements, with atomic numbers from 84 to 92, are slowly changing into lead (atomic number 82, mass number 207). This change is because of the phenomenon of radioactive decay (see RADIOACTIVITY).

Radioactive elements are unstable and gradually emit particles and energy, changing through a series of isotopes (see ISOTOPES) of decreasing mass until stable, nonradioactive lead is reached. For example, uranium-238 goes through 14 stages of decay, which involve the gas radon-222, lead-214, and lead-210. The rate of change at each stage varies greatly. For example, half a gram of uranium-238 will have changed into thorium-234 after 5 billion years; this period is known as the half-life. The half-life of lead-214 is less than a half hour, while that of lead-210 is 22 years.

Knowledge of half-lives can be used to date geological specimens. For example, by comparing the concentrations of uranium-238 and lead-206 in rocks, it is possible to work out that Earth is at least 4.6 billion years old.

A similar technique is based on the decay of radon-222 to lead-210. The gas radon escapes from rocks into the air and decays rapidly; the lead then falls to the ground and is taken up by ice or other surface materials. This method can be used for dating ancient rocks, glaciers, and even recent sediments.

A CLOSER LOOK

See also: COMPOUNDS AND MIXTURES; METALS.

Further reading:
Atkins, P. W. *The Periodic Kingdom*. New York: Basic Books, 1995.

LEPTONS

Leptons are a family of elementary particles. They appear to be simple, pointlike particles with no internal structure. This means that no experiments done thus far have given any evidence of leptons having a measurable diameter or of being composed of smaller parts. The behavior of leptons in magnetic fields shows that they have an intrinsic angular momentum (or spin) of one-half. Experiments conducted with particle accelerators show that leptons do not experience the strong force (the short-range force that holds protons and neutrons together in atomic nuclei).

Discovery of the positron

In the early 1930s, the atomic structure of all matter was explained by just three particles: protons, neutrons, and electrons. Protons and neutrons form the nucleus, the center of the atom, around which swirl a cloud of orbital electrons.

The simple, three-particle picture of atomic structure was disrupted by the discovery of a new particle in 1932. Carl Anderson, a physicist at the California Institute of Technology, was investigating cosmic rays coming from outer space. He was using a cloud chamber, an instrument in which the tracks of rapidly moving particles can be seen and photographed. When the chamber is placed into a magnetic field, positive and negative particles deflect in opposite directions. Anderson found some tracks that looked like normal electron tracks, but they curved the wrong way. He concluded that they must be made by positive electrons, which he named positrons.

The discovery of positrons was not unexpected. Four years earlier, English physicist P. A. M. Dirac (1902–1984), while trying to develop a form of quantum theory consistent with Einstein's theory of relativity, discovered an equation that described the behavior of electrons exactly—but it also described the behavior of another set of particles, which turned out to be the electron antiparticles, or positrons. The predicted particles would have the same mass as the electron but a positive charge. Anderson's discovery of positrons confirmed Dirac's prediction. Antiparticles for the proton and neutron were verified by later experiments in the 1950s. The existence of antiparticles for all subatomic particles has now been firmly established.

The discovery of the muon

In 1937, another new particle was discovered in cosmic rays. This particle produced tracks in a cloud chamber that curved considerably less than those of electrons or positrons. Less curvature indicates a more massive particle, and the new particle was shown to be heavier than an electron but lighter than a proton. It was named muon because mu (μ) is the Greek letter m, standing for medium mass. Both positive and negative muons were found, created from primary cosmic rays in Earth's upper atmosphere. Muons can be viewed as a heavier version of the electron and positron. In fact, negative muons can orbit a positive nucleus in the same way as electrons. They emit a pattern of wavelengths similar to that of the hydrogen atom, which consists of one proton and one electron. Muons and electrons have a so-called weak interaction (which is about 10^{12} times weaker than the strong interaction) with protons and neutrons.

This false-color bubble-chamber photograph shows the symmetrical production of lepton particles, electrons (green) and positrons, or antielectrons (red).

CONNECTIONS

● Experiments carried out with **PARTICLE ACCELERATORS** have helped to improve our understanding of leptons.

● Leptons are one of three families of **SUBATOMIC** particles.

CORE FACTS

■ Leptons are a family of elementary particles.

■ Leptons are simple, pointlike particles with no internal structure and a spin of a half; they interact weakly with each other and with other particles.

■ The lepton family of particles has six members: electrons, muons, tauons, and three different neutrinos.

■ Muons can decay radioactively into electrons.

■ All particles with a half integer spin, such as leptons, are called fermions.

SIX MEMBERS OF THE LEPTON FAMILY

Particle	Mass	Charge	Antiparticle
Electron (e^-)	1	(-)	positron (e^+)
Electron neutrino (ν_e)	0	0	antineutrino ($\overline{\nu}_e$)
Muon (μ^-)	207	(-)	antimuon (μ^+)
Muon neutrino (ν_μ)	0	0	antineutrino ($\overline{\nu}_\mu$)
Tauon (τ^-)	3500	(-)	antitauon (τ^+)
Tau neutrino (ν_τ)	0	0	antineutrino ($\overline{\nu}_\tau$)

Masses are given relative to the electron.

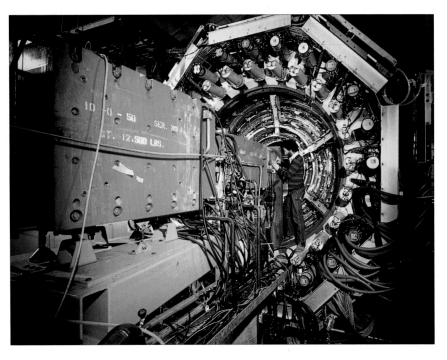

Using the Mark I detector at the Stanford Linear Accelerator Center (SLAC) in California, scientists have made numerous discoveries, including that of the tau lepton in 1975.

BOSONS

All particles are either fermions or bosons. Fermions are elementary particles with a half integer spin. They include leptons, quarks (see QUARKS), and baryons. Bosons are particles that have a spin of zero or any whole number.

Electromagnetic interactions, such as the repulsion between two electrons, can be represented by an exchange of particles called photons between the particles. Photons have a spin of zero and therefore are bosons. Gravitational interactions, such as the attraction between the Sun and Earth, are also attributed to the exchange of a boson called a graviton. The nuclear strong attraction that exists between quarks is due to a third type of boson, which has been given the whimsical name gluon. Finally, for the weak interaction, the exchange boson particle can be positive, negative, or neutral, and is designated by W^+, W^-, and Z^0.

Experimental evidence for these bosons was obtained with the high energy particle accelerator at the European Center for Nuclear Research in Switzerland in the 1980s.

neutral, it would leave no track in a cloud chamber. It was named neutrino, meaning "little neutral one." Its symbol is the Greek letter nu (ν) because it sounds like the start of the word *neutrino*.

Neutrinos are very difficult to detect directly. However, they can collide with protons, causing a rare secondary reaction. By setting up a special detector near a nuclear reactor (where neutrino production is abundant), the existence of neutrinos was verified experimentally 25 years later by Frederick Reines (Nobel Prize co-winner in 1995). In this reaction, a neutrino collided with a neutron, producing a proton and an electron:

$$\nu_e + \text{neutron} \longrightarrow e^- + \text{proton}$$

The fact that neutrinos do exist was experimentally established in 1962 by a team of scientists at the Brookhaven National Laboratory. These scientists also established that the neutrino associated with the electron differs from the neutrino associated with the muon, a discovery that brought them the Nobel Prize for physics in 1988.

Muons decay by radioactivity into electrons, accompanied by the emission of two distinctly different neutrinos. One is called the electron neutrino (ν_e) and the other is the muon neutrino (ν_μ).

In the 1970s, an accelerator was built at Stanford University in which beams of electrons and positrons could be made to collide. In such head-on collisions, a large amount of energy is concentrated into a very small space, allowing the formation of new particles (see PARTICLE ACCELERATORS). Among these was a superheavy electron-like particle, which was named tauon (τ). The tauon can be positive or negative, having a mass almost twice that of the proton. It decays by radioactivity into the muon and presumably also forms a tau neutrino (ν_τ), although that has not so far been verified experimentally.

A major goal of modern physics is to identify the particles that are truly fundamental—that is, particles that are not composed of other simpler particles. It is believed that all particles having rest mass other than that of the leptons are composed of quarks. There are six different basic types of quarks, which physicists refer to as flavors. Quarks also have a second property that physicists call color. The fact that there are six different basic kinds of quarks, which also appear to be pointlike and have spin quantum number one-half, and six different kinds of leptons, has led some physicists to conjecture that the leptons are simply the same six quarks appearing in another color, and so are part of the same family.

H. GRAETZER

See also: ELECTRONS AND POSITRONS; PARTICLE ACCELERATORS; PARTICLE PHYSICS; QUARKS.

Further reading:
Lederman, L. *The God Particle*. New York: Dell Publishing, 1994.
Sutton, C. *Spaceship Neutrino*. New York: Cambridge University Press, 1992.

Muons and electrons belong to the lepton family of particles. Another lepton was suggested theoretically in 1931 by Wolfgang Pauli (1900–1958). He was trying to resolve a puzzle in radioactivity in which beta particles are emitted with less energy than predicted. Pauli suggested that the missing energy might be carried off by an undetected neutral particle. Being

LIGHT

Light is a form of electromagnetic radiation. It occupies part of the spectrum of many kinds of electromagnetic radiation, including radio waves, microwaves, X rays, and gamma rays (see ELECTROMAGNETIC SPECTRUM). What gives each form of radiation its special character is its wavelength. The wavelength is related to the amount of energy each kind of radiation carries and affects the way it interacts with matter. What we call visible light is a small range of wavelengths from about 380 to 750 nanometers of the electromagnetic spectrum.

We call this light visible because human eyes and the eyes of most other living creatures can detect only these particular wavelengths. These wavelengths also happen to be the ones to which the atmosphere of Earth is most transparent, so most living creatures seem to have adapted to use the kind of radiation that is most available.

Our eyes respond differently to different wavelengths within the visible range, and we see these differences as color (see COLOR). So, light with shorter wavelengths—around 350 nanometers—is seen by most humans as violet, and light with longer wavelengths—around 750 nanometers—is seen as red, with a range of other colors in between. A mixture of all wavelengths of visible light that come from the Sun is seen as white. An absence of light is seen as black.

Radiation with wavelengths slightly longer than visible light is called infrared light. It is also invisible to human eyes, but we can feel it as heat.

Studying light

The study of light is called optics (see OPTICS). This is divided into geometrical optics, which deals with the behavior of light—how it moves and travels—and physical optics, which deals with the nature of light.

In geometrical optics, it is often convenient to speak of "rays" of light. A light ray is simply an imaginary straight line along the path that the light follows. When a light ray strikes a surface, it may be absorbed, transmitted, or reflected. If all light striking an object is absorbed, none is reflected to our eyes, and the object appears black. If the object absorbs some wavelengths and reflects others, the "color" of the object is the color our eyes perceive of the mixture of wavelengths that are reflected.

A rainbow is formed when white light from the Sun is refracted and reflected in raindrops, mist, or spray and is split into the visible spectrum of colors.

If most light passes through a material without interruption, so that we can see objects from the other side, the material is said to be transparent. If the light passes through but is scattered on the way, the material is called translucent. If a transparent or translucent material absorbs some wavelengths but not others, white light shone through the material will emerge on the other side as colored light.

When light is reflected from a completely smooth surface, it bounces off in accordance with the law of reflection (see REFLECTION AND REFRACTION): the angle of incidence equals the angle of reflection. The angle of incidence is the angle an approaching light ray makes with a line drawn perpendicular to the surface. The angle of reflection is the angle the departing ray makes with the perpendicular line.

CORE FACTS

- Visible light is that part of the electromagnetic spectrum—ranging from about 380 to 750 nanometers—that can be detected by the human eye.
- Passing white light through a triangular glass prism causes the refracted light to separate into bands of different colors.
- Light travels in small parcels of energy called photons.

CONNECTIONS

- During the **DAY** our major source of light is the **SUN**.

- Fluorescent materials absorb **ULTRAVIOLET RADIATION** and release their **ENERGY** as visible light.

- Electric lightbulbs lose a great deal of energy as **HEAT**.

This narrow yellow laser beam illustrates how light is reflected by a flat, smooth surface and obeys the law of reflection.

Interference

Throughout the 17th and 18th centuries, scientists debated whether light was composed of waves, as sound is, or of some sort of tiny particles. Those who accepted the particle theory pointed out that light apparently travels in straight lines, and when it strikes a smooth surface it bounces off, just as a billiard ball bounces off the side of the table. The most important evidence for the wave theory was the phenomenon of interference (see DIFFRACTION AND INTERFERENCE).

Suppose a beam of light is split into two parts and sent over slightly different paths, so that one path is slightly longer than the other. When the two parts of the beam are brought back together, they may "interfere," either reinforcing or canceling each other. Although light waves do not look like the waves on the surface of water, for example, they behave in much the same way. If two waves come together in such a way that the trough of one wave matches the peak of the other, the two waves cancel out, producing darkness. If they are brought together in such a way that the peaks and troughs match, they reinforce one another. When a series of light rays whose waves vary in the way they mix are spread across a flat surface, they alternately cancel and reinforce. The result is a series of light and dark bands called interference fringes.

Diffraction

If light were composed of particles and traveled in straight lines, then the shadow of a sharp edge should be perfectly sharp. The fact that it is not was another objection to the particle theory. When the shadow of a sharp edge is examined very closely, it appears that some light has "leaked" around the edge. When the area beyond where the edge of the shadow should be is examined even more closely, interference fringes are found. The explanation for this is that each point on a wavefront is actually radiating light in all directions; but waves radiating in any direction other than straight ahead interfere with, and are canceled by, waves from other points on the wavefront. When an object gets in the way, light from the parts behind it cannot be part of the mix, so some of the light passing by near the edge travels off at an angle into the shadow area. Light from each point near the edge travels a slightly different distance to where the shadow falls, causing interference fringes.

Refraction

When light passes at an angle from one transparent medium, such as air, into another, such as water or glass, it bends. This is easily seen when you push a pencil into a glass of water. The pencil appears to be crooked, because the light coming from the part of the pencil that is underwater is bent as it moves from the water to the air.

Refraction occurs because light travels more slowly in glass or water than in air. In simple terms, when light strikes a piece of glass or the surface of water perpendicularly, it passes through. But if it strikes it at an angle, part of its wavefront hits first and slows down first. This causes the wavefront to turn, just as a line of skaters holding hands would turn if the skaters at one end slowed down before the others. The amount by which the light is bent is governed by the refractive index, which depends on the difference in the speed of light in the two materials.

The angle a light ray passing into the second material makes with a line perpendicular to the surface is called the angle of refraction. When light passes from a medium in which it travels faster, such as air, to one in which it travels slower, such as glass or water, the angle of refraction will be less than the angle of incidence.

When light passes from a dense material to a less dense one, the angle of refraction will be greater than the angle of incidence. If light in this situation strikes at a very great angle, the angle of refraction will be so steep that the light ray is bent back into the slower medium; in other words, the light is reflected rather than refracted. This is called total internal reflection. You can see this when swimming underwater: if you look at the surface from a very great angle, the surface looks like a mirror. Total internal reflection is also what makes it possible for glass fibers to carry light long distances without much loss of intensity (see FIBER OPTICS).

A prism is a piece of glass (or sometimes quartz) shaped so that the faces will refract light in certain useful ways. Some prisms utilize total internal reflection to bend light at right angles; they are used in periscopes and other optical instruments. Others can turn images upside down. Different wavelengths of light are refracted by different amounts, so when white light is passed through a trianglar prism it is separated into a series of colored bands, each emerging at a slightly different angle.

Wave versus particle

In the mid-1600s, English physicist Sir Isaac Newton (1642–1727) argued that light was composed of particles, while Dutch mathematician, physicist, and astronomer Christiaan Huygens (1629–1695), who had been doing experiments on interference phenomena, argued that it was composed of waves. Newton was the foremost physicist of the time, and most other scientists followed his lead. However, in the early 1800s, Englishman Thomas Young (1773–1829) performed experiments that clearly demonstrated interference, and Frenchman Augustin Fresnel (1788–1827) continued this work about a decade later. These experiments finally compelled acceptance of the wave theory.

It was not until the early 20th century that scientists began to see that light somehow behaved as if it were composed of both waves and particles. Light is now thought to travel in tiny packets of energy called photons. A photon has no mass, but it has a specific energy. It behaves like a particle flying through space: when it strikes an object that it cannot penetrate, it bounces off and the light is reflected. But each photon also has the characteristics of a wave. It has a wavelength, and the wavelength is related to the amount of energy the photon carries. The intensity, or bright-

ness, of light is a measure of the number of photons present, not of the energy of each photon.

An atom is composed of a positively charged nucleus surrounded by negatively charged electrons. The electrons can reside only at certain energy levels. Electrons can move from one energy level to another, gaining or losing energy. When an electron drops from a higher to a lower energy level, the amount of energy given off is always exactly the same for a particular jump. It is different for different kinds of atoms but always the same in the same kind of atoms.

For an electron to move to a higher energy level, the atom must absorb exactly as much energy as that electron would have given off in moving down from the same level. Thus atoms can absorb only photons with certain energies. Since the energy of a photon is related to its wavelength, some materials absorb light of some colors and not others.

When electrons are raised to a higher energy level than the one they normally occupy, the atom is said to be excited. Electrons in an excited atom have a natural tendency to drop back down to their usual energy level, releasing energy. Often this energy is released in the form of another photon, and the wavelength of the photon is determined by the size of the electron's jump. This is why most materials, when given extra energy by heating or some other means, give off light of specific colors.

Light sources

Every known light source generates light in the same way. Atoms absorb energy when the substance they make up is heated, when an electric current is passed through it, when they undergo chemical or nuclear reactions, or when they absorb photons. This raises

When sunlight strikes the surface of a body of water, some of the light may be reflected and some may be transmitted into the water. The light that is transmitted into the water may be refracted, or bent, as it enters the water.

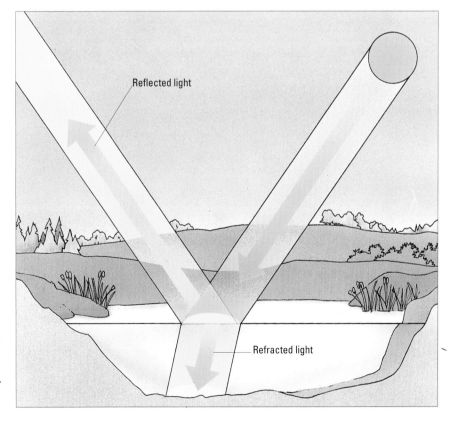

Reflected light

Refracted light

French physicist Augustin-Jean Fresnel is known for his work in optics, and his studies of interference were instrumental in the acceptance of the wave theory of light.

electrons in the atoms to higher energy levels. Eventually these electrons fall back to lower levels, and the atoms release energy in the form of photons.

In our most important source of light, the Sun, nuclear fusion supplies energy that heats everything around it to a hot gas, which in turn radiates light in the visible, infrared, and ultraviolet ranges, as well as radiating radio waves, X rays, and gamma rays.

In the common incandescent lightbulb, an electric current is passed through a thin wire to heat it "white hot," meaning that its atoms are emitting photons from several different electron energy levels, generating a range of wavelengths of light. Incandescent bulbs are not very efficient. As little as 10 percent of the electrical energy they use may be converted to light, with the rest being given off as heat.

Early in this century, it was discovered that a high-voltage electric charge passed through certain gases would cause the atoms of gas to emit light. The first gas used in this way was neon, so signs and decorations using gas-discharge light are popularly called neon signs, although today they use many different gases and mixtures of gases to obtain a variety of colors.

Some atoms and molecules absorb photons in the ultraviolet range and release their energy as visible light. Substances containing these atoms or molecules are said to be fluorescent. When a current is passed through mercury vapor, much of the light given off is in the ultraviolet range. A gas discharge tube is filled with mercury vapor and coated on the inside with a material that fluoresces in ultraviolet light. This produces light in the visible range; this is the principle of a fluorescent lamp.

Some substances convert energy released from chemical reactions into visible light. These substances are called phosphorescent. Some living things generate light by phosphorescence, including fish that use phosphorescent light to find their way in the depths of the ocean and the familiar firefly. So-called cold light devices sold at carnivals and novelty stores also use chemical phosphorescence.

PHOTOCHEMISTRY

All chemical reactions involve changes in the electrostatic attraction between atoms. When an atom absorbs a photon and raises one of its electrons to a higher energy level, the arrangement of electrons can be changed enough to cause a chemical reaction, either separating one or more atoms from a molecule or making atoms more likely to bond together. The study of how light can influence chemical reactions is called photochemistry.

The most familiar example of photochemistry in action is photosynthesis, the process by which molecules of chlorophyll in green plants use light to accomplish a complex series of reactions that change water and carbon dioxide into carbohydrates.

Another example is photography, where light, striking molecules of silver compounds in the emulsion of photographic film, makes these molecules able to react with chemicals in the developing solution, releasing metallic silver to form dark areas in the negative.

Light can also cause small molecules to join together to form polymers (see POLYMERS), a process called photopolymerization. In

photopolymerization, it is not necessary for light to strike all the molecules; one photon absorbed by one molecule may release a molecular fragment that starts a chain reaction. This property is used in plastics that remain soft until exposed to bright light, such as the photoresists used in preparing semiconductor chips.

Ultraviolet light from the Sun causes several chemical reactions in the atmosphere, including the production of ozone from oxygen (see OZONE LAYER). At high levels of the stratosphere, ozone helps to filter out ultraviolet light and protect living things on Earth from the damage it can cause. Unfortunately, other photochemical reactions with human-made pollutants in the atmosphere (principally chlorofluorocarbons) can reduce the amount of ozone.

Ozone is also produced in the atmosphere at sea level following photochemical reactions with other pollutants. Such reactions include the production of ozone from automobile exhaust and the reaction between hydrocarbons and oxides of nitrogen in sunlight. Ozone at sea level is undesirable and is one of the compounds in smog.

THE LIGHTBULB

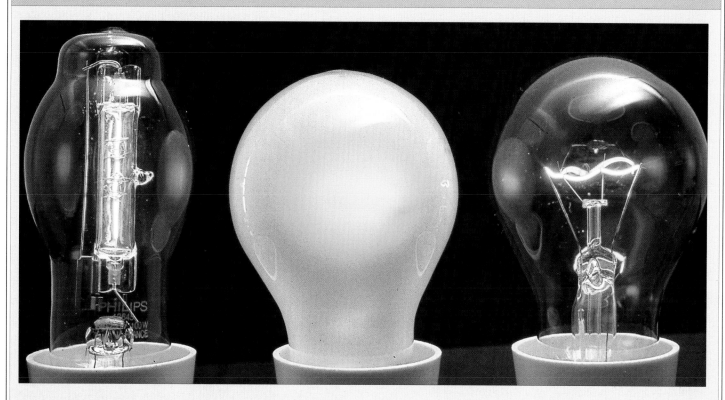

Thomas Alva Edison (1847–1931) is popularly credited with inventing the incandescent lightbulb (the word *incandescent* basically means "heated to the point of glowing"), but the idea was already an old one when Edison began working on it.

In 1802, British scientist Humphry Davy (1778–1829) showed how heating a platinum wire by passing an electric current through it would cause it to give off light. But the heat speeded up oxidation with the surrounding air and the wire quickly burned up. The use of an incandescent wire as a source of light had to wait for the development of the vacuum pump.

In 1845, American J. W. Starr obtained an English patent on a lamp using a heated carbon rod in a vacuum. Around 1860, English scientist Joseph Wilson Swan (1828–1914) expanded on Starr's work, using a carbonized (partially burned) paper strip as the filament. Edison began work on the problem in 1877, experimenting with nearly a thousand different filament materials, and on October 21, 1879, he successfully tested a lamp using a filament of carbonized cotton thread. Meanwhile, Swan had also produced a practical lightbulb. Edison lost a patent-infringement suit to Swan, and the two eventually went into business together to market the lamp.

Carbon filament lamps were used until 1908, when a process was developed for drawing fine filaments of tungsten. Tungsten filaments in a bulb filled with a mixture of argon and nitrogen burn brighter and last longer than carbon filaments.

What Edison did to earn his place in history was to engineer a practical lamp that could be mass-produced and to develop a system for generating and distributing electric power, which he marketed widely to cities.

HISTORY OF SCIENCE

The most modern light source, the laser (see LASERS AND MASERS), depends on the fact that when certain atoms in an excited state encounter a photon of the same frequency the atoms are prepared to emit, the atoms emit that photon in the same direction as the stimulating photon.

In a laser, a light-emitting material is "pumped" by a source of energy such as an electrical discharge or another light source. The photons emitted from its atoms are trapped between two reflecting surfaces, causing them to bounce back and forth, striking more and more atoms and causing them in turn to emit more photons. This produces a narrow beam of light that is monochromatic, meaning it is composed of light of the same wavelength, and coherent, so that the peaks and troughs of the waves are in step and are said to be in phase.

Velocity of light

Until the 17th century it was thought that light traveled instantaneously from place to place. Italian astronomer Galileo Galilei (1564–1642) was the first to suggest otherwise. He suggested a crude method of measuring the velocity of light by having two experimenters time distant flashes of light. There is no record that he ever tried it, but others who attempted it later, over a distance of about a mile, could see no time interval.

In 1675, Danish astronomer Ole Rømer (1644–1710) noticed that the time taken by one of Jupiter's moons to orbit the planet varied depending on where Earth was in its orbit. He realized that this was because light from Jupiter took longer to reach Earth over a longer distance. Using the value then known for the diameter of Earth's orbit, he calculated

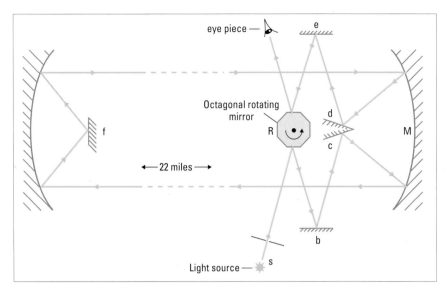

In 1926, U.S. physicist Albert Michelson used the eight-sided rotating mirror system to determine the speed of light with a high degree of accuracy.

POLARIZED LIGHT

The wave motion of a photon traveling through space is actually an oscillation of the energy of the photon back and forth between an electric and a magnetic field. These fields are at right angles to one another, and both are perpendicular to the direction in which the photon is traveling.

In most cases, the electric and magnetic fields of the photons in a beam of light are arranged in random directions. When they are all in the same direction, the light is said to be polarized. Light can become polarized when it is reflected from a shiny surface or when it is scattered by molecules of a gas or by fine dust particles. Examples are the light reflected from the surface of water or light in a clear blue sky.

Light can also be polarized when it passes through certain crystals and chemical solutions. The Polaroid Corporation, best known today for its instant cameras, first manufactured sheets of plastic material that could polarize light, using methods invented by Dr. Edwin H. Land (1909–1991). Sheet polarizers are used in sunglasses, photographic filters, and scientific instruments.

All polarizers, including reflecting surfaces, work by absorbing all light except that with an electric field aligned in one direction. Thus, polarized light will pass easily through a polarizer with which it is aligned, but it will be stopped by one aligned at right angles. Two polarizing sheets will appear transparent when aligned with one another, but they will appear black if one is rotated 90 degrees. Polarizing sunglasses make use of this property to cut glare reflected from the surface of water or smooth pavements; polarizing filters over a camera lens are used to darken the sky without changing its color.

Polarization may be linear, circular, or elliptical. In linear polarization, the direction and strength of the electric field remain constant. In circular polarization, the direction of the field is rotated around the direction of travel as the photon moves, but the strength of the field is still constant. In elliptical polarization, the direction of the field rotates, and its strength oscillates between high and low points with each rotation.

Certain molecules in solution will circularly polarize light passing through them. Some of these molecules exist in two forms, called stereoisomers (see ISOMERS; STEREOCHEMISTRY). Stereoisomers consist of the same atoms, arranged in the same shape, but with the position of some atoms reversed, making "right-handed" and "left-handed" versions of the shape. In some cases, one stereoisomer of a molecule will circularly polarize light with the direction of the electric field rotating clockwise, while the other stereoisomer will cause the field to rotate counterclockwise. This property can be used to find out which stereoisomer is present, even if the chemical properties of the two forms are the same.

A CLOSER LOOK

the velocity of light to be about 140,000 miles per second (225,000 km/s).

In 1849, French scientist Armand Fizeau (1819–1896) made the first laboratory measurement, sending a beam of light through a rotating toothed wheel to a mirror 5 miles (8 km) away. He adjusted the rate of rotation of the wheel so that a pulse passing between two teeth would return just in time to pass through the next notch. From the distance and the speed of the wheel he calculated the speed of light to be 158,100 to 213,900 miles per second (255,000 to 345,000 km/s).

A year later, another Frenchman, Jean Foucault (1819–1868), performed a similar measurement using a rotating mirror; in the time it took light to travel to a distant mirror and back, the rotating mirror turned just far enough to deflect the beam a few hundredths of an inch. This resulted in a value of 185,177 miles per second (298,000 km/s).

American physicist Albert Michelson (1852–1931) later refined this method, using an eight-sided mirror and measuring over a distance of 22 miles (35 km), eventually arriving at a velocity of 186,280 miles per second (299,774 km/s).

The development of microwave interference and the atomic clock made it possible to determine the velocity of light by measuring the wavelength of a beam of radiation of known frequency. Since the frequency is the number of waves passing a point in a second, the frequency multiplied by the wavelength gives the speed. Measurements by various methods fall between 299,792.458 and 299,792.60 km/s.

In 1957, the speed of light was defined by international agreement as $186,282.42 \pm 2$ miles per second ($299,792.5 \pm 4$ km/s). All earlier attempts to measure the speed of light depended on finding the time taken by light to travel a measured distance. Today the speed of light is a defined quantity, and that quantity is used to measure distance. The "standard meter" is defined as the distance traveled by light in $\frac{1}{299,792,458}$ th of a second.

The speed of light is also a fundamental constant of the Universe, represented in mathematics by the letter c. It enters into formulas that do not at first seem to have anything to do with radiation traveling through space, such as Einstein's famous equation $E = mc^2$, which gives the amount of energy released when a certain mass of matter, m, is converted to energy, E.

W. STEELE

See also: COLOR; DIFFRACTION AND INTERFERENCE; ELECTROMAGNETIC SPECTRUM; FIBER OPTICS; OPTICS; REFLECTION AND REFRACTION.

Further reading:
Falk, D. S., Brill, D. R., and Stork, D. G. *Seeing the Light: Optics in Nature, Photography, Color, Vision, and Holography*. New York: Harper and Row, 1986.
Giancoli, D. C. *Physics: Principles with Applications*. Englewood Cliffs: Prentice-Hall, 1991.

LIGHTNING AND THUNDER

Lightning is a powerful electrical discharge and thunder is the acoustic shock wave that accompanies it

In the time it takes you to read this sentence, lightning will strike Earth over 200 times. Lightning is static electricity, the same kind of shock you get when you touch a doorknob on a dry day, only much bigger. It occurs when electrical charges become so concentrated that they surge through the air in a brilliant electrical discharge. In general, lightning develops during unstable weather conditions, when warm, moist air rises from Earth's surface to form the towering cumulonimbus clouds associated with thunderstorms (see CLOUDS). Lightning can travel from a cloud to Earth, from a cloud to the surrounding air, within a single cloud, or between clouds.

Electrification of clouds

The basis of lightning is a process called charge separation. This occurs when positive and negative charges that usually exist together are pulled apart. According to one theory, clouds become electrified when hailstones fall through an area of supercooled ice crystals. As liquid droplets collide with a hailstone, they freeze and latent heat (see LATENT HEAT) is released. This keeps the surface of the hailstone warmer than that of the surrounding ice crystals. When a warmer hailstone collides with a colder ice crystal, a net transfer of positive ions occurs between the hailstone and the ice crystals (this transfer occurs from the warmer object to the colder object). This makes the hailstone negatively charged and the ice crystals positively charged.

Charge separation takes place in cumulonimbus clouds as a result of strong vertical air currents. These currents carry the negative charge to the bottom of the cloud, making the top side of the cloud positively charged (see CLOUDS). Consequently, there are regions of strong negative and positive charge within the cloud. Since opposite charges attract, the charges in these regions are inclined to rush together. However, charge can only flow along a conductive electrical path.

Clouds are not the only place where electrical charge concentrates during a thunderstorm. The

strong negative charge at the base of a thundercloud results in a deficit of electrons at the surface of Earth, just beneath the cloud. This is called inductive charging, and it is also responsible for a phenomenon called St. Elmo's fire (see the box on page 672).

Eventually, when the charge imbalance between the clouds and Earth becomes large enough, lightning strikes. Although lightning between and within clouds is most common, the lightning that travels from a cloud to Earth is more important to humans and has been studied extensively. High-speed cameras have helped reveal that the lightning we see as a single flash occurs in distinct stages.

Lightning can occur between two adjacent clouds if the difference in electrical charge between them is larger than that between either of the clouds and the ground.

Lightning stages

A stroke of lightning from a cloud to the ground begins when the electrical field near the base of the cloud is sufficiently large to strip electrons from the atoms in the surrounding air. The electrons rush toward the base of the cloud and then toward the ground along a conductive path called a leader, a column of ionized air about 4 in (10 cm) wide and 160 ft (50 m) long. When one leader ends, there is a brief pause before another bursts forward. Because the ionization path progresses in short, jagged steps, it is called a stepped leader.

As a stepped leader approaches the ground, a region of positive charge, called a streamer, moves up into the air through any conducting object, such as a tree or building. When one of these streamers contacts with a downward-moving stepped leader, large numbers of electrons flow into the ground and a larger, bright return stroke carries a strong positive charge upward into the cloud.

The first return stroke is usually followed by others, each preceded by a dart leader that reionizes the

CONNECTIONS

● High-speed **PHOTOGRAPHY** has helped reveal that the lightning we see as a single flash occurs in distinct stages.

● Lightning is a pulse of **ELECTRICITY** that flows to balance the difference in electrical charges in the **ATMOSPHERE**.

CORE FACTS

■ Lightning is an electrical discharge in the sky; it usually develops when warm, moist air rises from Earth's surface to form cumulonimbus clouds.

■ Before lightning occurs, a process called charge separation takes place in clouds. Negative charge tends to migrate toward the base of the cloud. This induces a positive charge near the ground. Eventually, there is a rush of current between the ground and the cloud.

■ There are different types of lightning. These include: forked, sheet, and ball lightning.

■ Thunder is a sound shock wave generated by lightning.

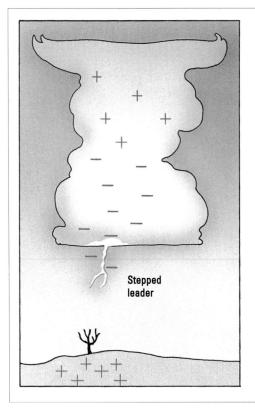

Stepped leader

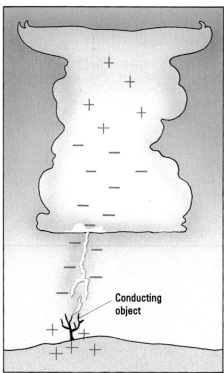

Conducting object

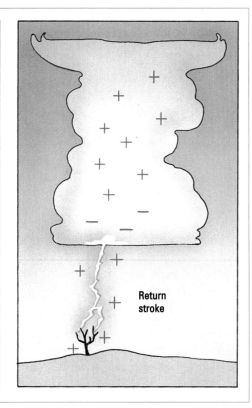

Return stroke

A stroke of lightning develops in three stages. First, when the negative charge near the bottom of the cloud becomes large enough to overcome air resistance, electrons rush toward Earth. Second, positive charge, called a streamer, moves up into the air through any conducting object, such as trees. Third, when the electrons meet the positive charge, a strong electric current carries the positive charge up into the cloud.

path taken by the step leader in one continuous motion. As many as 25 strokes may travel through the original channel. These strokes happen so quickly, one after the other, that they are seen as a single flash. A typical lightning flash involves a potential difference between the cloud and the ground of 100 million volts.

Types of lightning

The most common form of lightning is forked lightning. This is made up of many branches, although a single branch carries most of the electricity. In high winds, the main conducting channel may be blown sideways, so that each return stroke travels along a slightly displaced path. The resulting flash is called ribbon lightning.

Lightning that occurs within a cloud or between clouds may not be seen directly but will light up the sky. This lightning is called sheet lightning because it is obscured from view as if it were behind a sheet.

A rare form of lightning is called ball lightning. This type occurs near ordinary lightning. It takes the form of a sphere, ranging in diameter from 1 in

(2.5 cm) to 10 ft (3 m), although 8 in (20 cm) is typical. White, yellow, orange, or red, ball lightning is not well understood.

Thunder

Lightning can generate an acoustic (sound) shock wave. During the return stroke, the large electrical current traveling through the conducting channel heats the surrounding air. This sudden intense heating causes an explosive expansion, which reaches the ear as a loud boom. Sometimes, just before the loud crack of nearby lightning, a faint click can be heard. Scientists think this is the sound of the stepped leader.

Although lightning is seen almost the instant it is produced, there may be a delay before the thunder is heard. This is because sound waves travel relatively slowly and take longer to arrive the farther away they are. Sound waves also dissipate more quickly in air than light does; in general, thunder cannot be heard beyond a distance of about 20 miles (30 km). This fact can be used to calculate the distance from a lightning strike. Since sound waves travel at about 1000 ft (300 m) per second, it takes 5 seconds for sound to travel 5000 ft (1500 m), roughly a mile (1.6 km). Thus, the number of seconds between the flash and the thunder, divided by five, is the distance from the strike in miles.

P. TESLER

See also: ATMOSPHERE; CLOUDS; ELECTRICITY.

Further reading:

Ahrens, C. D. *Meteorology Today: An Introduction to Weather.* St. Paul: West Publishing Company, 1994.
Holton, J. R. *An Introduction to Dynamic Meteorology.* San Diego: Academic Press, 1992.

ST. ELMO'S FIRE

St. Elmo's fire is not really a fire at all but a glow sometimes seen on the masts of ships, the wings of airplanes, and other pointed objects during electrical storms. This glow is usually green, blue, or purple and is accompanied by a sizzling or crackling noise, due to an electrical discharge. This discharge occurs because strongly charged clouds overhead induce a charge on the object, which becomes very concentrated on the pointed parts. When this charge becomes sufficiently concentrated, it leaks out and there is a discharge. In earlier times, Mediterranean sailors thought that the glow at the tip of their masts was a sign that St. Elmo, the patron saint of sailors, was watching over them.

LIMESTONE

Limestone is a chemical sedimentary rock composed mainly of calcium carbonate

A rock is called limestone when its composition is more than 50 percent calcium carbonate ($CaCO_3$). This usually takes the form of the mineral calcite, with smaller amounts of aragonite (also $CaCO_3$, but with a different crystal structure) and perhaps dolomite (calcium magnesium carbonate, $CaMg(CO_3)_2$). Limestone may sometimes also contain minor amounts of quartz (sand grains), clay, iron minerals, and organic matter.

Limestone differs from terrigenous sedimentary rocks (those made of material derived from the land by erosion) in several important ways (see SEDIMENTARY ROCKS). It is formed by the chemical or biochemical precipitation of calcium carbonate from seawater or groundwater (see GROUNDWATER), and its formation is usually strongly influenced (both in the fixation of calcium carbonate and in the processes of deposition and lithification) by once-living creatures. It is also very susceptible to postdepositional chemical and textural alteration.

Limestone can be variable in appearance but is usually white or gray. The presence of impurities can make it yellow, brown, dark gray, or black. Many limestones contain abundant fossil remains.

How limestone is formed

The fundamental process in the formation of limestone is the precipitation of calcium carbonate from seawater or groundwater, and the most important way in which this occurs is through the activity of living organisms. Corals, diatoms, and various shelled creatures such as mollusks, echinoderms, and brachiopods all secrete calcium carbonate in the form of a shell or skeleton. When these creatures die, their soft parts decay, but the shells remain to provide the raw material for limestone. The shells can be deposited intact or be broken up by the action of waves or other living creatures to produce finer-grained sediment and carbonate mud (called micrite). Under the right conditions of temperature and pressure, calcium carbonate may also be precipitated inorganically, as in caves (see CAVES).

The limestone rock face revealed in this quarry in Germany was deposited in the Jurassic period, 208 to 144 million years ago, and was formed by the agglomeration of calcareous shells of various organisms. Such limestones are used for building materials or as a gravel aggregate.

Types of limestones

Limestones display a great variety of rock textures that reflect a wide range of depositional environments. These have been described using various highly technical classification systems that usually require the examination of thin sections of the rock under a microscope. Some of the more common varieties that can be distinguished in the field are described below.

Skeletal limestones are composed mainly of the shells of once-living creatures and are usually named after the predominant fossil type—for example, coralline limestone and crinoid limestone. Grain size is variable, ranging from the size of the largest shell down to microscopic fragments. Chalk is a special type of skeletal limestone composed of the calcareous parts of minute planktonic organisms (those that float passively in a body of water) that were deposited on the ocean floor far away from any source of terrigenous sediment.

Oolitic limestone is composed of oolites cemented together with calcite. Oolites are spherical to subspherical fine to coarse grains (0.1 to 1.0 mm in diameter) of carbonate sand formed in warm, shallow water agitated by wave action, where accretion of calcium carbonate has taken place around a nucleus—for example, a sand grain or a tiny shell fragment.

Microcrystalline limestone, also called micrite, is a very fine grained (less than 0.04 mm in diameter) limestone formed from carbonate mud deposited in

---

CORE FACTS

■ Limestone is formed by the chemical or biochemical precipitation of calcium carbonate from seawater or groundwater.

■ Most limestone deposits are composed of the remains of once-living organisms.

■ Limestone is easily dissolved in acidified rainwater, and in wet climates it forms a distinctive landscape called karst topography.

■ Limestone is an economically important natural resource; it is used in many ways in the construction industry, in agriculture, and in industrial processes.

CONNECTIONS

● Many chalk deposits were laid down during the **CRETACEOUS PERIOD**.

● Some limestone **ROCKS** contain impurities of **CLAY** and **SAND**.

LIMESTONE AND FOSSILS

Limestones occur in all continents and in strata of all ages from Archaean to Quaternary, and they are still forming today. They can form in fresh water or salt water, but the vast majority (over 99 percent) are marine in origin, and they account for around 20 percent of Earth's sedimentary rocks. Much of our knowledge of the evolution of marine creatures has been obtained from the study of fossils preserved in limestone (see FOSSIL RECORD).

The shallow marine environments in which many carbonate sediments are formed have always had a rich and varied fauna, so limestones are an ideal place to look for fossil remains. Most carbonate sediments were created by the activity of marine organisms, and the fossil remains of these creatures are the main constituent of many limestones. These fossil remains are often well preserved because they have been deposited fairly rapidly after the organisms died, although some will have been eroded and broken up by the action of waves and currents. In some places (such as the Permian limestones of Texas and Louisiana), ancient coral reefs have been preserved, allowing geologists to study whole communities of fossil species.

A CLOSER LOOK

Calcite is a typical sedimentary mineral and is one of the most important constituents of limestone.

a sheltered environment such as a lagoon or lake. Travertine is an inorganic limestone, formed by the evaporation of calcium carbonate-rich fresh water in rivers, springs, and caves, and deposited by certain hot springs in volcanic regions.

Another important carbonate rock is dolostone, which is composed entirely of the mineral dolomite, calcium magnesium carbonate, $CaMg(CO_3)_2$. Geologists are still debating the details of how dolostones are formed, but most are thought to be the result of the postdepositional alteration of limestone.

Limestone in the landscape

Calcite in limestone is soluble in groundwater that has been acidified by decaying plant matter or dissolved carbon dioxide (which combines with water to form carbonic acid: $H_2O + CO_2 \rightarrow H_2CO_3$). As the water percolates through joints, faults, and bedding planes, it dissolves the limestone, enlarging the fissures to create caves and caverns (see CAVES).

Karst topography is the distinctive landscape developed by the action of groundwater on limestone. (It is named after the Karst limestone plateau on the Italian-Slovenian border near Trieste.) The best-known karst regions in the United States are in Florida, Kentucky, and southern Indiana.

Karst regions are characterized by a lack of surface drainage, because any rainfall quickly finds its way underground. The roofs of caverns eventually collapse to form sinkholes, which are enlarged over time and merge into each other to produce solution valleys, with springs and disappearing streams. In humid, tropical climates, the rapid dissolution of limestone by heavy rainfall creates a spectacular landscape of hills called tower karst, as seen in southern China.

Uses of limestone

Limestone and its metamorphic equivalent, marble (composed of a coarsely cystalline, interlocking network of calcite grains), are important building stones and are especially prized as decorative stones. One of the best-known decorative limestones in the United States is the oolitic limestone from Bedford, Indiana, which was used in the construction of the Empire State Building in New York. Fine-grained limestones are crushed and used as aggregates for building highways and railroads.

When limestone is heated in a kiln, carbon dioxide evaporates to leave calcium oxide, better known as lime or quicklime. Lime is widely used in agriculture to improve acidic soils, and it is an important constituent of cement, plaster, and mortar. Lime and limestone are also used in making steel, glass, and ceramics.

Porous limestones can act as underground reservoirs for groundwater, oil, and natural gas. The limestones of the Jurassic and Cretaceous periods in Saudi Arabia and Iran form the reservoir rocks for the world's largest known petroleum deposits. Limestone is also an important host rock for ore deposits, including lead, zinc, silver, and fluorite (calcium fluoride, CaF_2).

N. WILSON

See also: CAVES; LANDFORMS; ROCKS; SEDIMENTARY ROCKS.

Further reading:

Prothero, D. P. and Schwab, F. *Sedimentary Geology.* New York: W. H. Freeman and Company, 1996.
Tucker, M. E. and Snow, V. P. *Carbonate Sedimentology.* Cambridge, Massachusetts: Blackwell Science, 1990.

LIPIDS

Lipids are substances, principally biological, that are insoluble in water but soluble in organic solvents

Lipids—from the Greek word *lipos*, meaning "fat"—occur widely in nature. They are not soluble in water but can be extracted from plant and animal cells with nonpolar organic solvents such as ether, chloroform, or benzene.

Lipids come in a variety of forms, including carboxylic acids (or fatty acids), glycerides (fats and oils), waxes, phospholipids, glycolipids, sulfolipids, terpenes, and steroids. Fatty acids are long-chain carboxylic acids. Examples include lauric acid, palmitic acid, and oleic acid. Only a small proportion of naturally occurring lipids consist of free carboxylic acids.

Glycerides (fats and oils)

Most of the carboxylic acids occurring in lipids are found as esters (organic compounds formed from the reaction of an acid and an alcohol) of glycerol. Glycerol is an alcohol with three hydroxyl (OH) groups, each of which combines with a carboxylic acid:

$$\text{glycerol} \quad \begin{array}{ccc} CH_2 & CH & CH_2 \\ | & | & | \\ OH & OH & OH \end{array}$$

Simple glycerides are esters in which all three hydroxyl groups of glycerol are esterified with the same carboxylic acid. Mixed glycerides are esters in which the three hydroxyl groups are esterified with two or three different acids.

The carboxylic acids that combine with these hydroxyls can be saturated—that is, with a full complement of hydrogen atoms attached to their carbon atoms—or unsaturated. With few exceptions, these carboxylic acids are straight-chain compounds, with from three to eighteen carbon atoms.

Glycerides that are solid at room temperature are called fats, whereas liquid glycerides are called oils. Glycerides with highly unsaturated fatty acid side chains have lower melting points than glycerides bearing fully saturated side chains and are therefore oils at room temperature.

Analysis

Because naturally occurring fats and oils are complex mixtures of glycerides, they do not have fixed physical and chemical properties. Even the melting point of a single type of fat can vary considerably, depend-

Beeswax, used to construct the honeycomb in beehives, is a well-known animal secretion that is composed of fatty acid esters.

ing on its source. There are a number of tests that are used to characterize fats.

The iodine number is a measure of the amount of unsaturated (double) bonds in the fatty acid component of a glyceride; it is the number of grams of iodine that will combine with 100 g of fat or oil. The saponification value is stated as the number of milligrams of potassium hydroxide (KOH) that are needed to hydrolyze (saponify) 1 g of fat; this is a measure of the total acid content of a fat or oil.

Saponification means "soap making," and this is exactly what treatment with KOH produces. The alkali reacts with the glyceride, releasing glycerol and forming salts of the long-chain fatty acids. Potassium soap, however, is too soft for domestic use, and ordinary soap is made with sodium hydroxide (NaOH).

Soap chemistry

Soaps are long chains of fatty acids with their carboxyl group attached to an atom of sodium or potassium. Unlike the bonds between the carbon atoms of the

CONNECTIONS

● Complex lipids are often composed of fatty **ACIDS** and other **FUNCTIONAL GROUPS**.

● A fat with a low melting point will be an **OIL** at room **TEMPERATURE**.

CORE FACTS

- Lipids are naturally occurring substances such as fats, oils, and waxes.
- Lipids are insoluble in water but soluble in organic solvents.
- Lipids play an important part in biochemistry and are found as phospholipids, glycolipids, and sulfolipids.

FATTY ACID CONTENT OF SOME FATS AND OILS

Fat or oil	Saturated acids (percentage)	Unsaturated acids (percentage)
Beef tallow	48–60	40–52
Butter	42–65	35–58
Coconut	74–99	1–26
Corn	9–16	84–91
Olive	8–27	73–92

CATALYTIC HYDROGENATION

Oils are glycerol esters of unsaturated fatty acids. If the saturation of the fatty acids is increased—that is, if hydrogen can be added to some of the carbon-carbon double bonds—the oils can be converted into solid or semisolid fats. In fact, this hydrogenation is very easily achieved, using nickel as a catalyst. The conditions under which the process is completed are relatively mild—temperatures under 392°F (200°C) and pressures between 20 and 40 lb/sq in—and the ester link with glycerol is not affected.

Beginning with cheap and plentiful oils, such as cottonseed, corn, and soybean, it is possible to produce fats that are closely related and have a similar consistency to butter or lard. For butter substitutes, the fat is usually mixed with slightly soured milk to give it a suitable flavor.

Long-chain alcohols can be obtained from oils and fats by hydrogenation at higher temperatures and high pressures, using copper chromite as a catalyst. This breaks the ester link, adding hydrogen to the –CO group. For example, lauryl alcohol, an important starting-point in the manufacture of detergents, can be made from coconut or palm kernel oil.

A CLOSER LOOK

chain, the bond between the carboxyl group and the alkali atom is ionic: the carboxyl group is negatively charged, and the alkali is positive—this end of the molecule is "polar."

In water, the soap molecules form micelles. The polar (carboxyl) end is water soluble, or hydrophilic (water-loving); the nonpolar (aliphatic) chains are hydrophobic (water-hating). The nonpolar chains cluster together, and the polar ends extend out to form a globular micelle. A soap cleans by attracting grease to the nonpolar end so that the grease can be dispersed (emulsified) in the water (see DETERGENTS).

Waxes

Waxes are found in plants, animal secretions, and the cell walls of some bacteria. The stored fat of marine animals contains a large quantity of waxes; beeswax, earwax, and carnauba wax, from a Brazilian palm, are other examples. Plant waxes also contain alkanes (see HYDROCARBONS).

Waxes differ from fats and oils in that they are not esters of gylcerol. Principally, waxy lipids are esters of long-chain fatty acids with long-chain alcohols with carbon numbers from 16 to 36; their molecular structure is therefore different from the triglyceride fats. A typical example is the wax found in sperm whale oil, which is an ester of palmitic acid ($C_{15}H_{31}COOH$) with cetyl alcohol ($C_{16}H_{33}OH$).

Sometimes the alcohol components of a wax are complex carbocyclic alcohols called sterols. Cholesterol is the best known of the sterols; it can be obtained by saponification from nearly all animal tissues. Plants contain the related sterols b-sitosterol, ergosterol, and stigmasterol.

Complex lipids (phospholipids, glycolipids, and sulfolipids)

Lipids that contain another functional group, as well as fatty acid ester groups, play a very important part in biological chemistry. This group can be a phosphate, a carbohydrate, or a sulfate, and the lipids are called phospholipids, glycolipids, or sulfolipids.

Esters of glycerol that contain two fatty acid chains and one phosphate group are called phosphoglycerides. They are found in the membranes of living cells. Here they make two layers of molecules, the nonpolar ends of the molecules attached tail-to-tail to form the membrane, and the polar phosphate groups projecting into the water on each side. Most of the membrane is made up of the fatty acid chains, so nonpolar molecules can dissolve in it, just as in a soap micelle, and pass through the membrane into the cell.

But how do polar molecules and ions get into and out of the cell? Embedded in the phospholipid layer are specific proteins that can identify and control the transfer of specific molecules and ions through the membrane into or out of the cell.

Glycolipids appear to play a similar role in cell chemistry. There are also complex lipids that do not contain glycerol but a long-chain amino alcohol called sphingosine. Derivatives of sphingosines are found in the brain and in nervous tissue, as well as in many plants and seeds.

Other lipids (terpenes and steroids)

If we take the general definition of lipids—that they are tissue constituents that can be extracted, not by water, but by various organic solvents—there is a wide variety of biochemical substances that answer this definition. They include terpenes and their derivatives (see TERPENES), which occur in many essential oils; steroids (see STEROIDS) and their related hormones; and vitamins A, D, E, and K.

B. INNES

See also: DETERGENTS; TERPENES.

Further reading:
Vance, D. E. and Vance, J. E. *Biochemistry of Lipids, Lipoproteins, and Membranes*. New York: Elsevier, 1991.

Coconut flesh can be pressed to produce coconut oil, which has a high percentage of saturated acids and is used in soap making and in other industrial processes.

LIQUID CRYSTALS

A liquid crystal is a substance that exists in a partially ordered stage between crystal and liquid

A liquid crystal is a material that does not simply pass from the solid state to the liquid state when heated. Instead, the liquid crystal has an intermediate, or mesomorphic, state in which the material exhibits some of the properties of a solid crystal and some of the properties of a liquid. Synthetic liquid crystals are important in devices such as digital watches and laptop computer screens. Natural liquid crystals are common in biological systems.

Liquid crystals were first discovered in the late 19th century, although early researchers studied them without understanding what they were;

German pathologist Rudolf Virchow (1821–1902), for example, noted that myelin, the lipid that sheathes nerves, seemed to have the optical properties of a solid even though it was a liquid.

A breakthrough in understanding

The first important step in understanding liquid crystals was made by German physicist Otto Lehmann (1855–1922), who built a device for his microscope that allowed him to control the temperature of objects he was studying. In 1888, Lehmann used the device to study the behavior of the compound cholesteryl benzoate, which had been brought to his attention by Austrian botanist Friedrich Reinitzer. Reinitzer had discovered that cholesteryl benzoate behaved oddly: when heated to 294°F (145.5°C), the solid form melted into a cloudy liquid, but when heated further, to 353.3°F (178.5°C), the liquid changed from cloudy to clear. Reinitzer concluded that the substance had two separate melting points.

Lehmann, studying the material with his microscope, observed that, in its cloudy liquid condition, it showed a characteristic of many solid crystals: it appeared to be able to rotate polarized light (see LIGHT). Lehmann suggested several names for this type of material, including soft crystals, floating

A polarized light micrograph of liquid crystals. The ability of liquid crystals to polarize light was one of the first properties to be discovered.

CORE FACTS

- Liquid crystals are mesomorphic—that is, at a particular temperature they can exist in a state between solid and liquid.
- In this state, the molecules are still held in place by molecular forces, but they can be moved together by mechanical, electric, or magnetic forces.
- Some liquid crystals can twist the plane of polarized light.
- Liquid crystals are important in the development of displays in digital watches and computer screens.

CONNECTIONS

- Liquid crystals can flow like a **LIQUID** but retain some of the properties of **SYMMETRY** of a crystalline **SOLID**.

- Cholesterol, a biological liquid crystal, is the starting point for many **STEROIDS**.

The graphics displayed on this computer screen are created by liquid crystals.

When the crystal is cool enough to be solid, the molecules line up in a particular direction and are kept in position by attractive forces between the molecules (see CRYSTALS AND CRYSTALLOGRAPHY). Heating the crystal adds energy to the molecules, and at a certain point they absorb enough energy to move out of their position—melting the crystal—but they still have insufficient energy to overcome the molecular forces that lined them up precisely. The molecules all remain oriented more or less in the same direction in space; this direction is called the director. On further heating, the molecules absorb enough energy for all order in the liquid crystal to be destroyed, and the crystal becomes fully liquid.

Nematic liquid crystals

In one common category, called nematic liquid crystals, molecules in the liquid are roughly parallel to one another, but the molecules do not form any detectable layers within the crystal. Early scientists used the word *nematic* because of something they observed when examining crystals under a microscope. The scientists would sandwich the crystal between two sheets of polarizing material, set in such a way that ordinarily the sandwich would look opaque. But the additional polarizing effect of the liquid crystals would allow some light to pass through the sandwich, producing an elaborate pattern of light and dark lines. This property, known as birefringence, is common with many solid crystals, and it was a key clue that molecules in a liquid crystal retained a precise orientation despite being a liquid.

In nematic crystals, the birefringence pattern is criss-crossed with threadlike lines. The word *nematic* derives from a Greek word meaning "thread."

Smectic liquid crystals

A second common category of liquid crystals is called smectic crystals. The crystal retains some vestiges of the original three-dimensional structure of the solid crystal even after it melts into the liquid crystal stage. A smectic liquid crystal is composed of distinct layers of molecules. The molecules in one layer are lined up in one direction, and the molecules in the adjacent layer may have a completely different orientation.

In one subcategory, known as smectic A, the director of each layer—that is, the direction in which the molecules are oriented—is perpendicular to the layer. In another category, called smectic C, the director is at an angle other than 90 degrees to the layer. The word *smectic* comes from the Greek word for "to clean," because smectic crystals are in some ways similar to soap films.

Cholesteric/chiral nematic liquid crystals

A third category of liquid crystals is known as either cholesteric or chiral nematic crystals. The former name comes from the fact that several cholesterol-containing compounds exhibit this characteristic. As

crystals, and crystalline fluids. Finally, he settled on the term *liquid crystals*, which is still in use today.

Types of liquid crystals

Liquid crystals exist in several varieties, according to the way in which the molecules are arranged (see the diagram below). For practical purposes, the molecules of most liquids can be thought of as being roughly spherical. However, the molecules in a liquid crystal are not spherical but almost rod-shaped.

This diagram shows the arrangement of molecules in an ordinary crystal (A), in a smectic liquid crystal (B), in a nematic liquid crystal (C), and in a liquid (D).

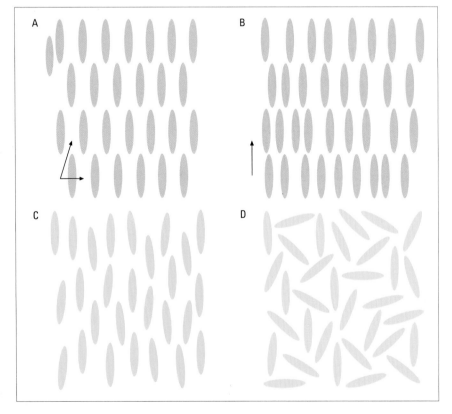

LIQUID CRYSTAL DISPLAYS

A liquid crystal display, or LCD, is a device that uses liquid crystalline materials to display information. LCDs are found in digital watches, calculators, and the flat panel displays used in portable computers, for example.

The first LCD was built in 1963 by researchers at the Radio Corporation of America. It made use of a property of some liquid crystal materials called dynamic scattering, in which a liquid crystal material is ordinarily transparent but becomes temporarily opaque while it is within an electric field. The electric field was created by sandwiching the liquid crystal between thin, transparent electrodes.

These dynamic scattering LCDs were an important first step, but they were soon abandoned when newer LCD technologies were developed that used less power and lasted longer. The next generation of LCDs was called twisted nematic displays.

The basic construction of this type of LCD is similar to that of the dynamic scattering LCD. It, too, sandwiches the crystal material between transparent electrodes that produce an electric field. But the inner surfaces of the electrodes are covered with a polymer (see POLYMERS), and that material is finely rubbed in one direction so that the crystal's director runs in that direction at the point where the crystal meets the electrode. The two plates have the direction of rubbing at right angles to each other, so that the crystal's director twists through 90 degrees from one plate to the other. Finally, polarizing material is placed on the outside of each piece of electrode, with the direction of polarization parallel to the direction of the rubbing on that plate's polymer.

When the electric field is off, light passes through the cell with no difficulty. The light is polarized by the material of the top electrode, has its polarity gradually twisted by the liquid crystal as it travels through the middle of the sandwich, and exits through the bottom electrode. But when the electric field is activated, the crystal temporarily untwists. Light still enters the top of the cell, but it is not gradually twisted perpendicularly by the liquid crystal, so the polarizing material at the bottom of the cell blocks the light, and the cell appears opaque or dark.

The electrodes can be cut into any shape. For example, an LCD capable of representing any digit from 0 to 9 can be formed by arranging seven dash-shaped segments as two squares on top of each other. Depending on which digit the unit is to represent, some sections of the device have the electric field activated and others have it deactivated.

A liquid crystal display (LCD) of the type used to show numbers. This type of display is familiar in calculators and digital watches.

Although twisted nematic LCDs are still common, a newer type is the supertwist nematic LCD. In a supertwist crystal, the director rotates through more than 90 degrees to 270 degrees. Supertwist LCDs can produce higher contrast in their displays than twisted nematic displays, an important consideration for designing computer screens that must display graphics and other data.

A key issue for engineers designing LCDs is the selection of the proper liquid crystal. Most LCDs use a mixture of several different crystals, chosen to create a useful set of characteristics, such as the range of temperatures at which it is mesomorphic. Another factor is how to control the hundreds of thousands of individual LCDs (pixels) that make up a computer screen or other display device. Each needs to be told precisely when to turn on or off. One solution has been the passive-matrix display, in which electric signals are sent inward from the horizontal and vertical sides of the display, timed so that they meet at the precise LCDs that need to be turned on at a given moment. In an active matrix device, each pixel is controlled by a transistor (see SEMICONDUCTORS) that acts as a switch. The resulting design is simpler than the passive matrix and also produces images that are sharper and have more vibrant colors.

SCIENCE AND SOCIETY

the latter name implies, this group is a subgroup of the nematic liquid crystals.

Chiral comes from the Greek word meaning "handedness." It can refer to left- or right-handedness. In these liquid crystals, the orientation of the molecules is not constant throughout the liquid crystal; rather, the average direction turns around at various depths of the crystal, forming a spiral or twist. The degree of twist is different at various temperatures, and the amount of twist also determines the color of light reflected from the crystals.

Physical properties of liquid crystals

The fact that molecules in a liquid crystal are not arranged randomly can influence other characteristics besides how it influences the transmission of light. For example, the arrangement of molecules

helps determine how the liquid crystal responds to mechanical stress on the crystal. If a shear stress is placed on a crystal—that is, a stress that tries to push the top half of the crystal off horizontally from the bottom half—the structure will yield less readily if the director is oriented vertically than if the director is oriented horizontally.

The structure of a liquid crystal also affects how the crystal reacts to electric and magnetic fields. If each molecule in the crystal has a small positive charge on one end and a small negative charge on the other—that is, it is polar (see POLARITY)—an external electric field will cause the molecule to orient itself in the direction of the electric field. For the crystal as a whole, the director becomes aligned with the direction of the electric field. Engineers use this characteristic in the design of liquid crystal displays.

LIQUID CRYSTALS IN THE BODY

Liquid crystals are not limited to watches and computer screens: they also play important roles in many biological systems, including the human body. One of the first liquid crystals ever to be studied by scientists was myelin, a fatty substance that acts as electrical insulation for nerve cells.

Liquid crystals are vital for the proper functioning of the membranes that sheathe every cell in your body. Cell membranes allow nutrients and other essential substances to enter the cell and waste products to be expelled from the cell. These membranes are double layers of lipids (see LIPIDS) that are in liquid crystal phase. The precise function of these liquid crystals is not yet known, but laboratory experiments suggest that lowering the temperature so that the lipid leaves its liquid crystal state causes the membrane to stop functioning properly.

Some scientists think that DNA, with its well-known helical shape, could be a cholesteric liquid crystal.

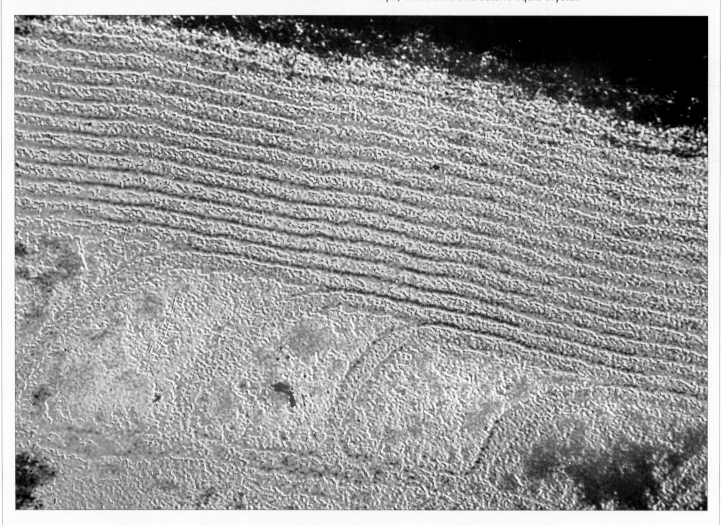

A false-color transmission electron micrograph showing the myelin sheath surrounding a nerve in the human ear. The layers of the myelin sheath are the orange bands at the top of the picture.

Similarly, when some liquid crystals are placed into an external magnetic field, that field can induce a magnetic dipole (see MAGNETISM) in the molecules. The molecules then become like tiny magnets that turn so that the dipole is aligned with the external magnetic field. In an electric field (see ELECTRICITY), the molecule generally turns so that its long side lies parallel to the electric field. But the magnetic dipole can be arranged either parallel to the long side of the molecule or at right angles to it: some liquid crystal molecules will turn so that they are parallel to the magnetic field, while others will line up at right angles to it.

V. KIERNAN

See also: CRYSTALS AND CRYSTALLOGRAPHY; LIQUIDS.

Further reading:
Collings, P. J. *Liquid Crystals: Nature's Delicate Phase of Matter.* Princeton, New Jersey: Princeton University Press, 1990.
Depp, S. W. and Webster, E. H. "Flat Panel Displays." *Scientific American*, **268**, pp.90-97, March 1993.
Hazen, R. M. and Trefil, H. *Science Matters*. New York: Doubleday, 1991.
O'Mara, W. C. *Liquid Crystal Flat Panel Displays: Manufacturing Science and Technology*. New York: Van Nostrand Reinhold, 1993.

LIQUIDS

Liquids are substances with a fixed volume but no fixed shape

Nearly all substances can exist in a solid, liquid, and gaseous state, and all matter consists of tiny particles, either individual atoms or molecules. When talking about liquids and gases, these particles are usually molecules.

Liquids differ from gases and solids in the way their molecules are packed. In gases, individual molecules are free to move about, so that a gas can be expanded or compressed over a very wide range. In a liquid, the molecules are so close together that they are virtually incompressible. The molecules, however, are still free to slip past one another. In a solid, molecules are so closely packed and tightly held that they can only vibrate in their fixed positions.

Three attractive forces

There are three main kinds of attractive forces between molecules in a liquid. Some molecules are polar: the bonds (see CHEMICAL BONDS) between the atoms making up the molecule are very strong, resulting in electrons being attracted toward one end of the molecule. This makes that end negatively charged and the other positively charged. The molecules tend to line up so that the negative end of one is close to the positive end of another. However, the attractive force is only about 1 percent of that of a covalent bond and decreases rapidly with distance.

When hydrogen is bonded to a small atom, such as oxygen, nitrogen, or fluorine, the molecule is strongly polar. Water (H_2O) is a good example. In water, the single electron of each hydrogen is attracted to the oxygen nucleus and joins the six electrons that surround it to form a covalent bond. This leaves two lone pairs of electrons around the oxygen nucleus that are not involved in bonding. These attract the hydrogen atoms of neighboring molecules. The force of this hydrogen bonding is five times stronger than ordinary polar attraction.

There is even an attractive force between nonpolar molecules. As electrons move around in a mole-

This high-speed photograph shows the impact of a water droplet on a pool of water. First the drop forms a crater, then the surface of the pool collapses, with the inward-rushing water colliding and rising to form a column.

CORE FACTS

- There are three main types of forces between molecules in a liquid: hydrogen bonds, London forces, and dipole attractions.
- When the rates of evaporation and condensation are equal in a solid/liquid system, the vapor molecules exert a vapor pressure on the liquid.

CONNECTIONS

● **METALS** can become liquid at high **TEMPERATURES**.

● The **MELTING AND BOILING POINTS** of **WATER** are higher than those of similar molecules, due to hydrogen bonding in the liquid state.

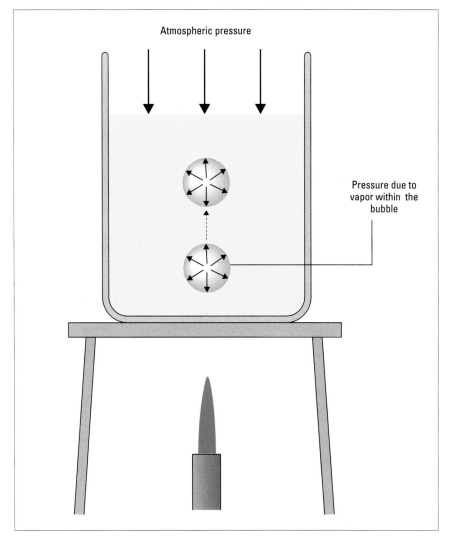

Atmospheric pressure

Pressure due to
vapor within the
bubble

*When a liquid boils, the
pressure of the vapor
within a bubble pushes the
liquid aside against the
opposing atmospheric
pressure.*

Physical properties of liquids

One of the most obvious properties of a liquid is the way it keeps its volume but not its shape. You can pour a glass of water into a pan: the amount and the space it takes up remain the same, but the water spreads into a different shape. A gas, on the other hand, will always expand to fill the container it occupies.

Another noticeable property of liquids is surface tension. At the surface of a liquid, the molecules are attracted only sideways and down; there are no molecules above them to balance this attraction. The result is that the surface tries to contract into a skin. Drops of water dripping from a tap will be drawn by this surface tension into a spherical shape; the water in an over-filled glass also tries to make a sphere—before the effect of gravity finally breaks it down (see CAPILLARITY AND SURFACE TENSION).

When a liquid has low surface tension—that is, when the attraction between the molecules is weak—it will flow easily over a solid surface and "wet" it. Gasoline, for example, is nonpolar and quickly spreads as a very thin film. But water, with its hydrogen bonding, has a high surface tension. When water is poured onto a greasy surface, the attractions between the water molecules and the surface are weaker than the hydrogen bonds within the liquid water, so the water forms droplets.

Vapor pressure

The temperature at which a solid turns into a liquid (its melting point) or a liquid into a gas (its boiling point) indicates the strength of the attractive forces between its molecules, although it is also related to the size of the molecules themselves. But even when a liquid is at a temperature well below its boiling point, it will gradually evaporate. The reason is that not all the molecules in the liquid have the same kinetic energy: some move much faster than others and escape from the liquid surface into the gaseous state, vapor. This reduces the average kinetic energy of the liquid; as a result, the liquid cools.

If the liquid partly fills a sealed container, some of the evaporated molecules will strike the walls of the container and give up some of their energy: they will condense and return to liquid state. When the rates of evaporation and condensation are the same, the vapor molecules exert a pressure in the container, above the liquid and upon it. This is called the vapor pressure. It increases with temperature until all the molecules have sufficient energy to escape, and the liquid boils.

E. KELLY

cule, there will be moments at which there is more electric charge on one side than on the other: the molecule will be momentarily polar and can be attracted to another molecule in the same state. It can also repel electrons in a neighboring molecule and so be attracted to the positive side of that molecule. These attractions are known as London forces, after German physicist Fritz London (1900–1954), who suggested their existence.

At high temperatures, metals and ionic compounds will also enter the liquid state. The forces between ions and atoms in ionic and metallic liquids are the same as in the corresponding solids, but the ions and atoms have too much kinetic energy above the melting point to be held in place by these forces.

See also: CAPILLARITY AND SURFACE TENSION; CHEMICAL BONDS; DIFFUSION AND OSMOSIS; FLUID MECHANICS; FORCES; GASES; MATTER; SOLIDS; WATER.

Further reading:
Giancoli, D. C. *Physics: Principles With Applications.* Englewood Cliffs, New Jersey: Prentice-Hall, 1991.

BROWNIAN MOTION

Robert Brown (1773–1858) was a distinguished Scottish botanist. In 1827, he was using his microscope to examine very fine pollen grains suspended in water, and he noticed that the grains constantly moved in what seemed to be a random way. In time, scientists realized that this "Brownian motion" was due to the movement of the water molecules that collided with the grains, causing them to recoil visibly. In 1905, physicist Albert Einstein (1879–1955) studied the experimental results and was able to calculate the approximate size and mass of atoms. He showed that the average diameter of an atom was of the order of 10^{-7} mm.

LUNAR MISSIONS

Spacecraft are sent to the Moon to accomplish a variety of scientific, engineering, and political objectives. As Earth's nearest celestial neighbor, the Moon represents a challenging yet attainable target for space probes (see MOON). In the earliest days of space exploration, the former Soviet Union and the United States tried to get spacecraft to the Moon simply as engineering tests, and both nations experienced failures before their first successful missions. The space race of the late 1950s and 1960s fueled a competition to accomplish impressive feats. As the programs matured, the missions became more sophisticated and accomplished ever more ambitious goals. After both nations decided to send humans to the Moon in the 1960s, some missions acted as precursors.

Sending a mission to the Moon presented exciting challenges to engineers and offered great rewards to scientists and everyone else interested in a greater perspective on the cosmos. It takes about three days for most spacecraft to travel from Earth to the Moon. Reaching the right point on the Moon (or in the vicinity of the Moon for fly-by and orbiting spacecraft) requires solving difficult navigation problems. In addition, once a spacecraft leaves the vicinity of Earth, it is outside the protective magnetosphere that shields the planet from solar and cosmic radiation. When spacecraft finally reached the surface of the Moon, they had to contend with extremes in temperature. During the two-week lunar day, the temperature can exceed 245°F (120°C), and the two-week lunar night can get as cold as -320°F (-160°C).

These orbital missions offered the opportunity to learn much about the Moon that could not be learned from Earth. Telescopes, with magnification limited by the turbulence of the atmosphere, could not see details on the Moon smaller than about a third of a mile (0.5 km). And because the Moon always keeps one face toward Earth, terrestrial observers could never observe the far side of the Moon. Obtaining a better view of the lunar surface was the goal of the initial missions to the Moon.

Later missions helped to determine the composition and distribution of the surface materials of the Moon, the structure of the interior, and more.

The first Soviet missions

Less than two years after its first space launch, the Soviet Union sent *Luna 1* toward the Moon. Launched on January 2, 1959, *Luna 1* was intended to hit the Moon as a demonstration to the world of the Soviet Union's superiority in space and as a test of the techniques involved in reaching this target 240,000 miles (386,000 km) away. It missed the Moon by over 3000 miles (over 5000 km), but its successor, *Luna 2*, did hit the Moon on September 13, 1959. *Luna 3*, launched on October 4, 1959, photographed 70 percent of the far side that humans had never seen. The film was developed on board and the *Luna 3* radioed back the stored images. The grainy photographs revealed a side different from the one familiar to Earth observers, with fewer of the large, dark, smooth areas known as the maria and with more mountainous regions, or highlands.

The first image of the far side of the Moon, shown above, was taken by the Soviet Luna 3 *spacecraft.*

CORE FACTS

- Robotic probes have successfully flown past the Moon, hit it, orbited it, landed on it, and returned lunar samples to Earth.
- In the 1960s and 1970s, the former Soviet Union and the United States took part in increasingly ambitious and productive lunar exploration missions.
- Most lunar exploration missions were conducted between 1959 and the early 1970s. After a hiatus of over 13 years, lunar exploration missions resumed in the 1990s.
- The first successful lunar probes were launched by the former Soviet Union, the United States, and Japan.

CONNECTIONS

- **SOVIET SPACE MISSIONS** to the **MOON** began before those of the United States.

- Investigations have shown that lunar **ROCKS** are similar to those found on **EARTH**.

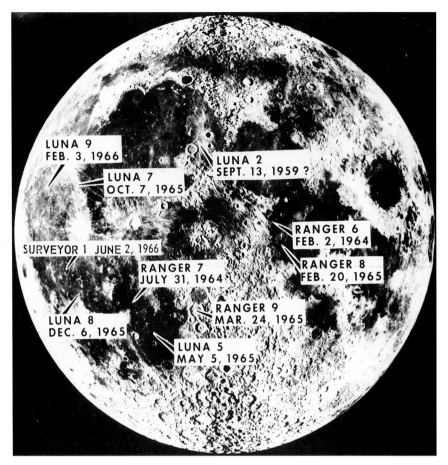

LUNA 9
FEB. 3, 1966

LUNA 7
OCT. 7, 1965

LUNA 2
SEPT. 13, 1959 ?

SURVEYOR 1 JUNE 2, 1966

RANGER 6
FEB. 2, 1964

RANGER 7
JULY 31, 1964

RANGER 8
FEB. 20, 1965

LUNA 8
DEC. 6, 1965

RANGER 9
MAR. 24, 1965

LUNA 5
MAY 5, 1965

The image above shows the impact points and landing sites of the first lunar unpiloted Soviet and U.S. spacecraft. The first soft landings on the Moon were achieved by the Soviet Luna 9 *and U.S.* Surveyor 1.

After these successes, the Luna probes suffered a series of failures through 1965, but exploration of the Moon was continued by some of the missions in the Zond ("probe") program. On July 20, 1965, *Zond 3*, a test model for future planetary probes, took 25 pictures of the far side of the Moon, including much of the area not photographed by *Luna 3*. Designed as precursors to piloted missions, *Zonds 5* through *8* orbited once around the Moon and returned capsules to Earth between 1968 and 1970. These probes carried plants and animals to test the effects of space travel on various organisms, and they took high-quality photographs during their journey.

Second-generation missions

Luna 9 became the first spacecraft to survive a landing on the Moon when it ejected a 23-in (58-cm) capsule one second before crashing on February 3, 1966. The nine pictures the capsule returned showed a smooth landing site with boulders near the horizon. This mission provided compelling evidence that the surface was not covered with thick dust, as some scientists had thought, and was not too fragile to support a spacecraft. Two months later, on April 3, *Luna 10* became the first spacecraft successively to orbit the Moon. Unlike the U.S. Lunar Orbiter spacecraft, it did not carry cameras, but its sensors showed that the Moon's magnetic field was much weaker than Earth's, if it existed at all, and that some of the surface was covered with basalt. It also searched for a lunar atmosphere. *Lunas 11*, *12*, and *14* made similar measurements from lunar orbit, with *Luna 12* also transmitting high-resolution photographs.

Completing an extremely successful year for the Soviet Union in lunar exploration, *Luna 13* landed on December 22 and measured properties of the regolith and took photographs.

All the Soviet Union's lunar probes came under the name *Luna* (except the cosmonaut-related Zond series and some unsuccessful missions given a Kosmos designation). The spacecraft designs changed over the years, but within each generation the spacecraft shared design features, with specific modifications for each type of mission. In contrast, the United States had three different programs with robotic probes. The first, the Rangers, were designed to crash into the Moon, radioing pictures throughout their descent. Following the conclusion of this program, Lunar Orbiters photographed the surface of the Moon from orbit, and Surveyors landed to make detailed investigations at a few select sites as precursors to the piloted Apollo missions.

U.S. robotic missions

The Ranger program began with two partially successful test missions in Earth orbit and then had four successive failures trying to reach the Moon. Nevertheless, after this series of disappointments, in 1964 and 1965, *Rangers 7*, *8*, and *9* returned a total of 17,255 views of the lunar surface, revealing features as small as 10 in (25 cm). These images showed details that were 1000 times smaller than those visible from Earth. By tracking the spacecraft from Earth, scientists greatly improved their estimate of the Moon's mass. When it crashed, each spacecraft created a crater about 36 ft (14 m) in diameter. Later missions returned photographs showing these impact sites.

During the Ranger program, the United States decided to focus its lunar missions on preparations for the landing of astronauts on the Moon. The five Lunar Orbiters, launched between August 1966 and August 1967, took photographs of 99 percent of the Moon, with emphasis on possible landing sites. *Lunar Orbiter 1* became the first U.S. spacecraft to orbit the Moon (or, for that matter, any celestial body other than Earth) and returned the first photographs from lunar orbit. All five Lunar Orbiters provided valuable experience in sending a craft into lunar orbit. Indeed, by precisely tracking their orbits, scientists further refined their determination of the mass of the Moon and learned that its gravitational field is highly irregular. Concentrations of mass under the surface, called mascons, perturbed the orbits of the spacecraft. This knowledge was useful in inferring the subsurface geology and in planning and operating subsequent lunar missions.

In parallel with the Lunar Orbiter program, the United States landed five Surveyor spacecraft on the Moon. Three days after its launch on May 30, 1966, *Surveyor 1* gently settled on the lunar surface. It corroborated evidence from *Luna 9* that the surface was firm enough to support a landing craft, bolstering plans for the landing of astronauts. Its cameras detected features as small as one-fiftieth of an inch

(0.5 mm), one million times smaller than could be seen from Earth. *Surveyors 2* and *4* were not successful, both crashing on the Moon, but the remainder of the program, concluding with *Surveyor 7*'s investigations in the rugged terrain near the large crater Tycho, was completely successful. By the final contact with *Surveyor 7* on February 21, 1968, the five landers had taken over 87,000 photographs, and some measured the composition and strength of rocks and soil at their landing sites. Scientists made a range of important discoveries from these measurements. Their investigations revealed that the lunar surface rocks were similar to terrestrial rocks, with silicates predominating. They also found that the Moon consisted of two different areas: a light-colored highland area and a dark-colored lowland. The highlands were made of minerals enriched with calcium and aluminum, and the maria (seas) were made of basaltic lavas enriched with iron and titanium. They found that the soils in these regions varied in composition, with fragments ranging from angular shapes to round shapes. They also varied in size, from diameters of the order of micrometers to centimeters.

Apollo

U.S. astronauts orbited the Moon on *Apollos 8* and *10* in 1968 and 1969 (see APOLLO MISSIONS), mainly to test techniques for the later landing missions; they also returned the highest-quality photographs yet taken from orbit. Except for *Apollo 13*, *Apollos 11* through *17* landed on the Moon between 1969 and 1972. On each mission, samples were collected and returned for analysis in laboratories on Earth, scientific instruments were distributed around the landing sites, and many photographs were taken. Readings from magnetometers confirmed the idea that the Moon did not have a molten core. However, the results did show that the Moon had a weak magnetic field when the rocks solidified billions of years ago, indicating that the Moon originally had a molten, iron-rich core. Seismometers set up by the astronauts also indicated that the Moon exhibits far less seismic activity than Earth. Only 3000 moonquakes were detected before the instruments were shut down, while there are hundreds of thousands of earthquakes. The moonquakes are also far weaker. Seismometers also showed that moonquakes are more frequent at a new moon and at a full moon. From this, geologists concluded that moonquakes are influenced by tidal forces.

In November 1969, *Apollo 12* astronauts landed only about 520 ft (160 m) from the long-inactive *Surveyor 3*, which had been on the Moon since April 1967. They returned portions of it to help determine how the harsh lunar environment affected the probe. The Apollo program returned 842 lb (382 kg) of lunar materials. These materials provided important information about the early history of the Solar System. In fact, all the samples taken were igneous rocks, indicating that the surface of the Moon was once molten. *Apollos 15* and *16* also each released a small satellite into lunar orbit before returning to Earth. *Apollos 15*, *16*, and *17* each had a surface vehicle (called Lunar Rovers) to help the astronauts traverse a larger area.

Third-generation Soviet missions

The Soviet Union's efforts to send cosmonauts to the Moon were not successful, but during the years that U.S. astronauts were landing on the Moon, the Soviet Union did continue its Luna missions with a new generation of spacecraft that were substantially larger and more capable than their predecessors. Between *Lunas 16* and *24*, only *Luna 18* and *Luna 23* did not complete their missions. *Luna 16*, which exceeded the size of the previous successful Soviet landers, executed a soft landing on September 20, 1970. Its robotic arm collected some soil, and the lower section of the spacecraft served as a launchpad for the upper part, so that it could begin its return trip to Earth.

A lunar rendezvous: astronaut Charles Conrad standing next to the Surveyor 3 spacecraft on the Moon. Conrad arrived on Apollo 12, and his lunar module can be seen on the horizon.

On September 24, 1970, *Luna 16* delivered its precious cargo of soil samples to Earth. The Soviet Union repeated this accomplishment with *Lunas 20* and *24*, returning a total of about 700 lb (320 kg) of lunar soil with the three missions.

The Soviet Union continued its orbital missions with *Luna 19* in 1971 and 1972 and *Luna 22* in 1974 and 1975. Instead of including a system to return samples to Earth on *Luna 17* and *Luna 21*, each landing stage carried a remote-controlled Lunokhod ("Moon rover") over 7 ft (2.1 m) across, equipped with cameras and sensors to study the soil. After its landing on November 17, 1970, *Lunokhod 1* spent 11 months investigating 6½ miles (10 km) of the lunar surface. *Luna 21* released *Lunokhod 2* on January 16, 1973, 110 miles (180 km) north of the Taurus-Lithrow region in Mare Serenitatis that the *Apollo 17* astronauts had explored the previous month. The two Lunokhod rovers returned over 100,000 photographs and made hundreds of measurements of lunar soil, including searches for a magnetic field and remote sensing of the chemical composition of the lunar surface.

Most Soviet missions landed on the eastern near side of the Moon, so scientists could only analyze the soils and rocks in this region. Their most notable work was on the Mare Crisium, a region with soils containing high concentrations of magnesium.

With changing political and scientific priorities, the number of missions to the Moon declined after the early 1970s. *Luna 24*'s mission to collect lunar soil samples in August 1976 was the last by any country for over 13 years. Some spacecraft flew by it for gravity assists; that is, to use the gravitational attraction of the Moon to change direction. Using the Moon's motion and gravity helps spacecraft to reach their destinations more efficiently and with less energy. Some spacecraft viewed the Moon on their way to other targets in the Solar System, but these missions were not geared specifically for the exploration of the Moon.

Revival of interest

In the 1990s, there was renewed interest in lunar missions. Japan became the third nation to send a spacecraft to the Moon with the launch of *Hiten* (named after a Buddhist angel that plays music in the heavens) in January 24, 1990. It was designed as an engineering test to help the country develop techniques in operating spacecraft in the vicinity of the Moon, and its primary mission was engaging in large Earth orbits with occasional passes by the Moon. During its first pass by the Moon, *Hiten* released a small daughter probe called *Hagomoro* (an angel's feather garment), which entered lunar orbit on March 19. A failed transmitter prevented *Hagomoro* from returning data. In February 1992, after its primary mission was complete, *Hiten* was itself sent into lunar orbit.

The United States' return to the Moon came on February 21, 1994, when *Clementine* entered orbit, over 21 years after the previous U.S. lunar exploration mission, *Apollo 17*, had concluded. Built to test a variety of sensors for defense against foreign missiles, the spacecraft trained its instruments on the lunar surface. With the improvements in technology that occurred during the previous decades, it collected high resolution photographs, measured the global topography, and even searched for ice (although none was found) in permanently shadowed craters, near the poles. *Clementine* data led to the first global color and elevation map of the Moon, and it was the first spacecraft to get clear views of the Moon's southern polar region, confirming the presence of the largest (1600 miles or 2500 km wide) and deepest (7½ miles or 12 km deep) impact crater known in the Solar System.

With the resurgence in interest in the Moon, other missions began to be developed. The U.S. *Lunar Prospector* is planned to make detailed maps of the composition of the lunar crust and the magnetic field and to study gases emanating from the Moon. Using experience from *Hiten*, Japan's *Lunar-A* is planned to make the first contact with the far side of the Moon. It would carry three penetrators that will be dropped from an orbiter. Two will be dropped on the near side and one on the far side, to form a network of seismometers and heat-flow probes and to permit the further study of the interior of the Moon. The penetrators will transmit their data to the orbiter for relay to Earth, so that even the penetrator on the far side, never visible from Earth, can contribute to solving the puzzle of the Moon.

M. RAYMAN

See also: APOLLO MISSIONS; MOON; SPACE.

Further reading:
Burgess, E. *Outpost on Apollo's Moon*. New York: Columbia University Press, 1993.
Wilhelms, D. E. *To a Rocky Moon: A Geologist's History of Lunar Exploration*. Tucson: The University of Arizona Press, 1993.

The Lunokhod ("Moon rover"), an unpiloted, remote-controlled Soviet lunar roving vehicle, provided many photographs of the lunar surface. The Lunokhod was driven around on the Moon's surface by radio commands from controllers on Earth.

MAGELLAN

Magellan is a robotic space probe that was launched to map the surface of Venus

On May 4, 1989, a space probe was launched from the U.S. space shuttle *Atlantis*. The probe was called *Magellan* and was named for Ferdinand Magellan (c.1480–1521), a Portuguese explorer who commanded the first circumnavigation of Earth (that is, a voyage completely around Earth) in the 16th century. The probe's main mission was to circumnavigate Venus from space and to use a radar device to map its surface, which is shielded from ordinary cameras by the dense clouds that fill its atmosphere.

Venus is a puzzle to scientists. In many ways, it could be regarded as Earth's twin: the two planets have roughly the same size, density, and gravity. But conditions on the two planets are sharply different: the atmosphere on Venus is composed of 96 percent carbon dioxide compared to 0.03 percent for Earth's atmosphere; this high carbon dioxide level produces a stronger greenhouse effect on Venus, so that trapped heat from the Sun brings the surface to a temperature as high as 887°F (475°C).

Early missions to Venus

Magellan was hardly the first spacecraft to visit Venus. Both the United States and the Soviet Union had been dispatching a number of spacecraft to the planet since 1961. They used two different strategies to try to pierce the Venusian clouds: bouncing radar waves off the planet's surface from orbit, as *Magellan* did, and landing a probe on the surface to study it directly.

The U.S. *Mariner 2* spacecraft flew past the planet in 1962 but did not land. In 1967, the Soviet Union's *Venera 4* descended on a parachute through the atmosphere of Venus, transmitting data on its composition over a period of 94 minutes, but radio transmission ceased at about the time of its landing on the surface. The first useful landing was accomplished in 1970 by the Soviet Union's *Venera 7*, which broadcast useful data on the atmospheric pressure and surface temperature from the planet's surface for 23 minutes. The first pictures of the surface were radioed back to Earth from the Soviet Union's *Venera 9* in 1975.

The Magellan spacecraft was launched from the cargo bay of the space shuttle Atlantis *on May 4, 1989.*

In 1978, the U.S. *Pioneer Venus Orbiter* mapped the planet's surface using a radar that was able to penetrate the atmosphere's thick clouds (see PIONEER). The Soviet Union's *Venera 15* and *16* missions, launched in 1983, had better radars with a resolution of 1¼ miles (2 km), about 50 times better than that of the *Pioneer Venus Orbiter*, but they mapped only a small portion of the planet.

These probes provided scientists with their first views of the overall topography of the planet. For example, the early Venus probes pinpointed three large regions of highlands. They also pinpointed mountain ranges, valleys, volcanoes, and impact craters. Data from the orbiters and landers suggested

CORE FACTS

- *Magellan*'s main mission was to map the surface of Venus from space; it was launched from the U.S. space shuttle *Atlantis* on May 4, 1989.
- *Magellan* used a sophisticated radar device that could penetrate the thick clouds in the Venusian atmosphere.
- Evidence from *Magellan* shows that the crust of Venus does have craters, faults, and volcanic eruptions, but they are not arranged in a way that suggests the crust exhibits plate tectonics.
- The probe burned up in Venus's atmosphere in October 1994.

CONNECTIONS

- **VENUS** has an **ATMOSPHERE** of hot and dense **CARBON DIOXIDE**.

This false-color perspective of three craters on the surface of Venus was constructed from data gathered by the Magellan radar-mapping spacecraft. The craters show radar-bright central peaks and are surrounded by radar-bright ejected material.

that Venus does have a crust, but it does not exhibit plate tectonics (see PLATE TECTONICS); this has set scientists debating about other mechanisms the planet uses to dissipate heat from its interior.

A near disaster

Magellan had some problems shortly after it arrived on Venus, which threatened to derail the entire mission. The problems were due, in part, to a design that cut costs: *Magellan* had only one antenna, which did two jobs—communicate with Earth and detect radar waves that were being reflected from the planet. While *Magellan* followed an orbit that carried it from the north pole to the south pole, it used the antenna to map the surface below. Then, while the satellite traveled back to the north pole, it would use the antenna to broadcast the radar readings back to Earth. But the satellite repeatedly malfunctioned, losing touch with Earth. After several months, engineers discovered that the problems were due to errors in the probe's computer programming, which were overcome so that *Magellan* could do its work.

Achievements of the Magellan program

By the end of its mission, *Magellan* had managed to map 98 percent of Venus's surface by piecing together many thin, north-south swaths of radar observations. These observations had a resolution that was 10 times better than those made by *Venera 15* and *16*. *Magellan* also gathered observations of the strength of the planet's gravity at various locations, resulting in a gravity map of the planet that would allow scientists to infer where masses of dense material might be present in the planet's interior.

One of *Magellan*'s earliest and most important discoveries was evidence that Venus definitely does

not have a system of tectonic plates like Earth's. Like earlier probes, *Magellan* spied volcanoes and faults, which on Earth would be evidence of two plates moving against one another. But plate tectonics on Earth also usually produces continent-spanning mountain ranges along the line where two plates are pushing against each other, and no extensive, linear mountain ranges were observed on Venus. *Magellan* also identified more than 900 of the craters pockmarking the planet's surface, but they seemed to be randomly scattered, and many seemed to have been undisturbed since they were formed. Many—though not all—scientists studying Venus think that massive volcanic eruptions may have wiped the planet's surface clean between 300 million and 500 million years ago and that the craters are no older than that.

Magellan measured the planet's gravity field from September 1992 to May 1993. During this period, it stopped using its radar system and broadcast a constant, unchanging radio signal back to Earth instead. If the satellite passed over a region of Venus with stronger gravity, the laws of orbital mechanics dictated that *Magellan* would inevitably speed up. This would create a Doppler effect (see DOPPLER EFFECT), slightly changing the frequency of the radio signal. Computers on Earth, knowing the details of the satellite's orbit, used the Doppler shifts to pinpoint changes in the gravity field of Venus.

Magellan also performed some dramatic stunts. In May 1993, after the radar was deactivated permanently, ground controllers used *Magellan* to test the feasibility of aerobraking—changing a satellite's orbit by making it skim the atmosphere slightly, producing friction. NASA engineers think that aerobraking could be used to make interplanetary probes less expensive. In September 1994, *Magellan* performed a windmill experiment, in which its solar panels were extended like the paddles of a windmill. The satellite dipped again into Venus's atmosphere, and ground controllers measured how much torque was required to keep the spacecraft from spinning.

That was only a prelude to *Magellan*'s end. In October 1994, the probe again extended its solar panels like a windmill and fired its rockets, but this time its orbit plunged deeply into the atmosphere. For about a day, the probe gathered data on the upper atmosphere and the probe's behavior in it. It inevitably burned up and was destroyed—but not before scientists got one last glimpse of the workings of Earth's twin.

V. KIERNAN

See also: MARINER PROBES; NASA; PIONEER; SOVIET SPACE MISSIONS; VENUS.

Further reading:

The Face of Venus: The Magellan Radar Mapping System. Edited by L. E. Roth and S. D. Wall. Washington, D.C.: National Aeronautics and Space Administration, 1995.
Morrison, D. *Exploring Planetary Worlds.* New York: Scientific American Library, 1993.

MAGMA

Magma is molten material that rises from Earth's interior and solidifies to form igneous rock

Overall, Earth's crust is composed largely of igneous rock; that is, rock formed from magma. Magma is a fluid mixture of molten rock, mineral grains, and dissolved gases. Once it surfaces, it is lava. When magma erupts as lava, it pours out onto Earth's surface and cools and solidifies to form extrusive igneous rocks, such as basalt and rhyolite. Magma that solidifies within the crust before reaching the surface forms intrusive igneous rocks, such as gabbro and granite (see IGNEOUS ROCKS).

The formation of magma

Most magma is formed at depths of about 60 to 180 miles (from 100 to 300 km) beneath the surface of Earth, in a zone of the upper mantle called the asthenosphere (see EARTH, STRUCTURE OF). In the asthenosphere, high temperature and pressure raise the mantle rock nearly to its melting point, so it is weak and has little resistance to shearing. This allows the movement of the tectonic plates that make up Earth's rigid outer layer (see PLATE TECTONICS).

Magma forms in localized regions, where conditions are favorable for the partial melting of the rocks of the upper mantle and crust. The specific details of magma formation depend on many factors including temperature, pressure, and the composition and water content of the rock.

Types of magma

The igneous rocks found in Earth's crust are formed from three main types of magma: rhyolitic, andesitic, and basaltic. The relatively less dense rhyolitic and andesitic magmas are associated with continental crust and the relatively denser basaltic magma with ocean crust. All magmas are composed mainly of silicon and oxygen (as silica, SiO_2). The amount of silica largely determines the type of magma and the way it behaves. Magmas low in silica can flow like a river, while those high in silica will ooze, more like spilled oatmeal. The most common magmas, basaltic magmas, contain the least amount of silica. They are

Lava channels form below the active vents of Mount Etna, Sicily.

CORE FACTS

- Magma comes from a zone in the upper mantle called the asthenosphere.
- Magma is a fluid containing molten rock, mineral grains, and dissolved gases; it mainly consists of silicon and oxygen.
- Magma flows onto Earth's surface in volcanic eruptions and solidifies to form igneous rock.
- Pockets of magma can be found near the boundaries of the tectonic plates, along the midocean ridges, and along the margins of continents.

CONNECTIONS

- **QUARTZ** is a common mineral found in **IGNEOUS ROCK**.

- Processes that separate remaining **LIQUID** from formed **CRYSTALS** in a cooling magma lead to the formation of different igneous rocks.

composed of 50 percent silica, a large proportion of iron and magnesium, and usually very little gas. Basaltic magmas have temperatures from 2010 to 2190°F (1100 to 1200°C). At the other extreme, rhyolitic magmas contain from 60 to 70 percent silica and usually large amounts of dissolved gases, mainly water vapor. These magmas are relatively cool, having temperatures of only about 1470°F (800°C). The temperature and properties of andesitic magmas are intermediate between those of the rhyolitic and basaltic magmas.

Tectonic setting

Much of Earth's volcanic activity occurs as basaltic eruptions along the midocean ridges, which are constructive plate boundaries where new ocean crust is generated as the lithospheric plates spread apart at rates of a few inches per year. Subduction zones around the rim of the Pacific Ocean are the other major volcanic region. Subduction zones are areas of compression between lithospheric plates where a more dense oceanic plate is pushed under a continental plate or another oceanic plate. Andesitic volcanoes occur in these zones where plates capped by oceanic crust plunge into the mantle and are remelted (see PLATE TECTONICS).

Oceanic volcanism

Eruption of basalt along the midocean ridges accounts for about 40 percent of the volcanism on Earth's surface. As the crust is pulled apart, the release of pressure causes partial melting of peridotite, the rock that makes up the mantle. The resulting magma, known as basaltic magma, flows into the rift above. Lavas that erupt onto the seafloor typically occur as rounded, lobate masses or pillows that have shiny black crusts of balsaltic glass, resulting from the rapid cooling of the lava by seawater. Some magma cools on the sides of the rift, with successive eruptions forming vertical layers called dikes.

Beneath the surface, the basaltic magma cools more slowly into the large-grained version of basalt, commonly called gabbro.

Continental volcanism

On the continents, volcanic activity can be basaltic, andesitic, or rhyolitic. Basaltic magma is more dense than the continental crust, which is andesitic in composition. Where the continental crust is pulled apart, as in fault rift basins, fluid basaltic magma can well up through the rifts to form extensive flood basalts. In places, basaltic magma causes partial melting of the continental crust and rhyolitic magmas are formed. The explosive eruptions of rhyolitic lavas form pyroclastic rocks such as frothy pumice, glassy obsidian, and tuff, made of volcanic ash.

Andesitic magmas are usually formed from the partial melting of the oceanic crust. This often takes place in subduction zones, where the lithosphere, with its covering of wet basalt, sinks into the asthenosphere. The presence of water in a magma can lower its melting point by several hundred degrees, so melting can occur at depths as shallow as 30 to 120 miles (50 to 80 km). Andesitic magmas can result from partial melting of the basaltic crust or by incorporating material from the continental crust as the magma rises to the surface.

J. FEDERIUK

See also: FAULTS; GEOLOGY; IGNEOUS ROCKS; LANDFORMS; PLATE TECTONICS; ROCKS; VOLCANOES.

Further reading:

Carroll, M. R. and Holloway, J. R. *Volatiles in Magmas.* Washington, D. C.: Mineralogical Society of America, 1994.
Hamblin, W. K. *Introduction to Physical Geology.* New York: Maxwell Macmillan International, 1994.
Physical Chemistry of Magmas. Edited by L. L. Perchuk and I. Kushiro. New York: Springer-Verlag, 1991.

This diagram shows how volcanic processes have spread across the Japanese islands. At depths between 60 and 90 miles (100 and 150 km), oceanic crust that has been carried downward by the descending lithosphere undergoes wet partial melting. Andesitic magma rises, creating volcanoes on the Japanese islands.

MAGNESIUM

Magnesium is a very light, low-density metallic element that is obtained from a number of ores and from seawater

Despite its name, magnesium is not magnetic—but there is a connection. It gets its name from the ore magnesia (MgO), which in turn gets its name from Magnesia, the district of ancient Greece in which the ore magnetite was also found (see IRON AND STEEL; MAGNETISM). The element was first isolated, by electrolysis of magnesia, by English chemist Sir Humphry Davy (1778–1829) in 1808, having been recognized by Joseph Black in 1755.

Comprising about 2.5 percent of Earth's crust, magnesium is the eighth most abundant element and the sixth most abundant metal on Earth. It is a member of the alkali earth metals that comprise group 2 (IIA) of the periodic table. Like all the elements in this group, magnesium is very reactive and is only found in nature combined with other elements.

Physical properties

Magnesium is a very light silvery-white metal, its density at 68°F (20°C) being only 1.738 g/cm³, and it is the lightest metal that can be used structurally. Its density is about a third less than that of aluminum (2.698) and about a fifth that of iron (7.873). When combined with other metals such as aluminum and zinc, magnesium produces very lightweight alloys that are, nevertheless, very strong. This makes magnesium alloys ideal for use in the manufacture of airplane and automobile parts, where strength must be maximized and weight minimized.

At 1204°F (651°C) and 2030°F ± 18°F (1110°C ± 10°C), the melting and boiling points of magnesium are relatively low, although its melting point is only about 18°F below that of another structural metal, aluminum. In some other respects, such as thermal and electrical conductivity, the properties of magnesium are also similar to those of aluminum.

Chemical properties

Magnesium is a very reactive metal—although not as reactive as the alkali metals of group 1 (IA) or the heavier elements of group 2. In general, the elements of group 2 are less reactive than those of group 1, because the latter need give up only one electron, compared with two for group 2 elements, in order to undergo a chemical reaction.

Magnesium, a very reactive metal, burns in air after ignition.

The elements in group 2 that are heavier than magnesium are calcium, strontium, barium, and radium. The valence electrons of these elements are more distant from their positively charged nuclei than is the case with magnesium. Thus, the electrons of these elements are more weakly held by their atoms, which increases their reactivity. The lower reactivity of magnesium is another reason this metal is well suited for structural use.

Nevertheless, magnesium is easily oxidized. For this reason, the metal is often stored under kerosene to protect it from exposure to air. When heated, it burns in air with a brilliant white flame, and in the early days of photography it was used to provide brief but intense illumination in flashbulbs. The finely divided metal is still used in pyrotechnics, including underwater flares and in blasting powders.

Magnesium reacts with most acids and nonmetals but tends not to react with alkaline substances. When heated, the metal is a very strong reducing agent. As such, it is used to free metals such as titanium, zirconium, and hafnium from their halide compounds. Under ordinary conditions, magnesium will not react readily with water, but, if the water is brought to a boil, it will react with it and liberate hydrogen gas.

Magnesium is an integral part of certain organic and biochemical compounds. For example, it is important part of chlorophyll, the compound that makes photosynthesis possible. Chlorophyll is a com-

CORE FACTS

- Magnesium is the eighth most abundant element on Earth and the sixth most abundant metal.
- Magnesium is very light and can be mixed with other metals to make alloys that are light and strong.
- Magnesium is a constituent of chlorophyll, so it is believed to be key to the evolution of life on Earth.
- Grignard reagents, a range of organic magnesium compounds, are important in many synthesis reactions.

CONNECTIONS

- **ELECTROLYSIS** of magnesium chloride provides free magnesium and chlorine **GAS**.

- Magnesium is used in **ALLOYS**, glass, **BATTERIES**, and many other things.

Symbol: Mg
Atomic number: 12
Atomic mass: 24.305
Isotopes:
  24 (78.7 percent),
  25 (10.13 percent),
  26 (11.17 percent)
Electronic shell structure: [Ne]3s²

VICTOR GRIGNARD AND HIS REAGENT

In the latter half of the 19th century, the field of organic chemistry—the chemistry of complex carbon compounds—was developing rapidly. Chemists were not only investigating the properties and structure of these compounds, they were attempting to synthesize ones that had never existed, as well as many that were relatively rare in nature.

The task was formidable and complicated. The reactions they thought would yield the expected products either could not be made to start or progressed too slowly to be practical. Shortly before the year 1900, French chemist François-Auguste-Victor Grignard (1871–1935) was trying to attach a methyl group (see FUNCTIONAL GROUPS) to a larger organic molecule.

Grignard needed a substance to catalyze (see CATALYSTS) the reaction. But what substance? He knew that another chemist had had some success synthesizing organic compounds by using as a catalyst a complex of zinc and an ether. This compound, however, did not work with methyl groups. The question remained, was there another metal that, when combined with an ether, would make his reaction work?

Grignard knew that small fragments of metallic magnesium could catalyze some organic reactions but not others. Perhaps, he thought, a mixture of magnesium and an ether would work. And so it did. The reaction that had eluded him now proceeded.

So were born a family of magnesium-based catalysts that are now used to synthesize an extremely wide variety of organic reactions. The Grignard reagents used in present-day laboratory work are alkyl magnesium halides, with the general formula RMgX, dissolved in a solution of ether. For his work, Grignard was awarded the Nobel Prize for chemistry in 1912.

DISCOVERERS

plex molecule constructed around a single atom of magnesium. Magnesium is also vital to the physiological processes of animals. Deficiencys can cause diabetes and high blood pressure. The daily dietary requirement for human beings is about 300 mg, which is easily obtained from a wide variety of foods.

Sources of magnesium

A major source of magnesium is seawater, in which the metal is found at a concentration of 0.13 percent. Magnesium is also found in abundance in various minerals, including magnesite (magnesium carbonate), dolomite (a double carbonate of calcium and magnesium), asbestos (a complex silicate), talc and olivine (other complex silicates), brucite (magnesium hydroxide), carnallite (a combination of potassium chloride and magnesium chloride), and kieserite (magnesium sulfate).

Magnesium minerals are mined throughout the world. For example, in the United States, commercial reserves of brucite, magnesite, and dolomite lie in California, Nevada, and Washington. Other countries in which magnesium ores are mined include Austria, Brazil, Canada, India, China, Russia, and Venezuela.

Manufacture

Magnesium is mass-produced by two distinct processes, electrolytic and silicothermic. The former is used to extract magnesium from seawater.

In the electrolytic process, the raw material is magnesium chloride. The steps are as follows:
1. Seawater is mixed with lime obtained by heating oyster shells (which are also recovered from the sea). This isolates magnesium chloride from other substances in seawater by converting the chloride into magnesium hydroxide:

$$MgCl_2 + CaO + H_2O \rightarrow Mg(OH)_2 + CaCl_2$$

2. The magnesium hydroxide precipitate is filtered off and then treated with hydrochloric acid to reconstitute soluble magnesium chloride, which is dissolved in water:

$$Mg(OH)_2 + 2HCl \rightarrow MgCl_2 + 2H_2O$$

3. This solution is evaporated, leaving magnesium chloride behind.
4. The magnesium chloride is then melted with a mixture of sodium and potassium chlorides and electrolyzed, producing metallic magnesium and chlorine gas. To prevent oxidation of the metal, carbon is sometimes added to the electrolytic cell.

The starting material for the silicothermic process, also called the Pidgeon process after its inventor L. M. Pidgeon of Canada, is the ore dolomite.
1. The dolomite is mixed with an alloy of iron and silicon called ferrosilicon.
2. The mixture is made into tiny bricks, which are heated in a furnace to 2192°F (1200°C).
3. The product is magnesium oxide (MgO), which is rapidly reduced by the silicon to a vapor of metallic magnesium.
4. The vapor is then cooled to produce crystals of magnesium metal, which are then melted and poured into molds to make ingots.

Magnesium compounds

Magnesium hydroxide is a white powder, only slightly soluble; it is mixed with water to make milk of magnesia. Magnesium salts are used as laxatives: magnesium sulfate is commonly known as Epsom salts. This compound is also mixed into fertilizers. In organic chemistry, the complex substances that magnesium forms with alkyl groups are known as Grignard reagents (see the box above) and are an important tool in the synthesis of alcohols and alkanes, for example.

C. PROUJAN

See also: ELEMENTS; METALS.

Further reading:
Barrett, J. *Understanding Inorganic Chemistry: The Underlying Principles.* New York: Ellis Horwood, 1991.
The Biological Chemistry of Magnesium. Edited by J. A. Cowan. New York: VCH, 1995.
Chemical Deicers and the Environment. Edited by F. M. D'Itri. Boca Raton: Lewis Publishers, 1992.

MAGNETIC POLES

The magnetic poles are two points near Earth's axis of rotation where the geomagnetic force is vertical

In ancient times, when humans looked up into the heavens at night, they saw one star that did not appear to move. This is Polaris—the North Star—and for hundreds of years, tracking its position was the only way that travelers had to find their way at night. The Chinese are believed to have been the first to use a basic magnetic compass (see MAGNETISM). This always pointed toward the North Star, even when there were thick clouds; it gradually dawned on people that Earth itself was a giant magnet, with north and south magnetic poles.

The magnetic poles are in the polar regions but do not coincide with the geographic North or South Poles—the ends of the axis about which Earth revolves—which are known as the "true" poles. Because of this, a compass does not point to true geographic north. In fact, the situation is even more complicated: because of variations in Earth's structure, the direction in which a compass does point varies, depending on where in the world you are. The acute angle between the direction the compass points and the geographic meridian through the same location is called the magnetic declination. Maps are made that show lines of equal declination, called isogonic lines.

At the magnetic poles, the geomagnetic lines of force are perpendicular to Earth. From the south magnetic pole, the lines arc up and outward, just like those of a bar magnet, going back down to Earth at the north magnetic pole. A free-moving magnetic needle will be parallel to Earth's surface only near the equator; elsewhere, it will dip toward the nearer pole. Directly above the magnetic poles, the needle will be vertical, with the north-seeking end of the needle pointing down above magnetic north and up above magnetic south. The angle that the needle makes with the horizontal is called the magnetic inclination. The magnetic inclination is positive when the north-seeking end of the needle dips below the horizontal and negative when it rises above the horizontal.

The magnetic field of force around Earth is called the magnetosphere. The magnetosphere is distorted by a stream of charged particles from the Sun called

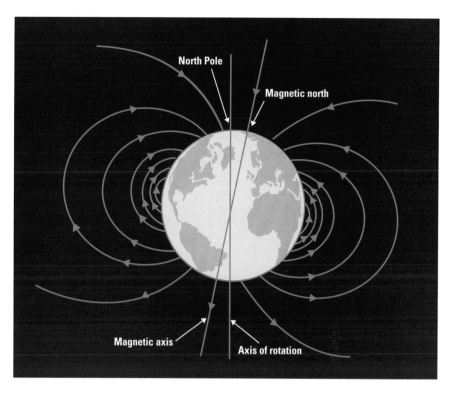

Earth has a magnetic field around it that has north and south magnetic poles, just like that of a bar magnet.

the solar wind (see SUN). The outer boundary of the magnetosphere, called the magnetopause, is much closer to Earth on the side of the Sun. On the opposite side it extends into a tail, making the magnetosphere look like a comet (see COMETS).

Magnetic polar regions
The center of the area identified as the north magnetic pole is at present at a latitude of about 73 degrees N and a longitude of about 102 degrees W. It is located in the Northwest Territories of Canada in the Parry Islands, part of the stark Arctic Archipelago between the Arctic Ocean (see ARCTIC OCEAN) and Baffin Bay. The south magnetic pole is just off the coast of Antarctica in the Indian Ocean (see INDIAN OCEAN) at a latitude of about 67 degrees S and a longitude of about 143 degrees E, which puts it just about at the Antarctic Circle near the Mertz glacier tongue (see ANTARCTICA).

Light shows at the poles
Pictures taken at night from satellites show a glowing auroral ring above the area around the geomagnetic poles in the night sky. From Earth, the auroras are luminous red and green rays, curtains, and arcs continuously changing in form and intensity at 60 to 600 miles (100 to 1000 km) above the horizon. Auroras are produced by the interaction of solar wind with the magnetopause. High-energy particles reach Earth from solar flares, creating disturbances in the magnetosphere that descend into Earth's upper atmosphere and excite oxygen atoms to emit green light and nitrogen to glow red.

CORE FACTS
- The north magnetic pole is located in Canada, some 800 miles (1250 km) from the geographic North Pole.
- The south magnetic pole is located at the edge of Antarctica, some 1600 miles (2550 km) from the geographic South Pole.
- A fossil record of the magnetic poles, found in the alignment of magnetic minerals in rocks, shows that the magnetic poles have reversed at intervals of less than a million years throughout Earth's history.
- The magnetic poles originate from the movements of the liquid core, deep within Earth.

CONNECTIONS

● Earth's **MAGNETISM** comes from the movement of its core, a mixture of metallic **LIQUIDS**.

● The south magnetic pole is off the coast of **ANTARCTICA** in the **INDIAN OCEAN.** The north pole is adjacent to the **ARCTIC OCEAN.**

The northern lights (aurora borealis) are the result of the interaction between Earth's magnetic field and the solar wind, an electrical conductor. The auroras are most dramatic around the magnetic poles.

Geodynamo and the magnetic poles

The magnetic poles originate deep within Earth. At Earth's center is a hot and dense solid inner core surrounded by a liquid outer core with a high temperature of 9000°F (5000°C). Both the inner and outer cores are rich in iron and nickel. Scientists explain geomagnetism by comparing the movement of the liquid core to a dynamo. Rotation of Earth, making convection currents within the liquid core, produce movements that can be compared to electricity flowing through a copper coil. This electricity flow produces magnetic force lines that emerge from one end of the coil and loop back through space to the other, forming a dipole (see MAGNETISM).

Magnetic poles and fossil magnetism

Over time, scientists have discovered that Earth's magnetic field is not static but changeable. Evidence for this comes from the orientation of magnetic minerals preserved in ancient lavas and sedimentary rocks. One feature of such magnetic minerals, such as hematite and magnetite, is that at very high temperatures they lose their magnetism, only to become magnetic again as they cool down. Therefore, when lava cools, the hematite and magnetite within it align themselves parallel to Earth's magnetic field at the time of their cooling. Similarly, when sediments containing magnetic minerals are laid down, they align as they become trapped in rock formations.

Studies of recently formed magnetic rocks seem to show that the position of the magnetic poles has wandered (during the last 7000 years) throughout a region within 11 degrees of the geographic poles. (In fact, we now know that the magnetic poles have not moved in relation to the continents at all, but the polar-wandering effect occurs because the continental landmasses themselves have shifted in relation to the poles. As a result, different continents have different polar wander curves; see CONTINENTS.) The record of Earth's magnetic field in the geologic past is called paleomagnetism.

Paleomagnetic studies have revealed another startling fact: the magnetic poles have reversed completely many times—at least 20 times in the last 10 million years alone. These studies have been useful in other areas of geology. For example, since magnetic reversals are irregular through time, reversal patterns form a unique fingerprint for each interval of geologic time and the rocks formed during those intervals. Studies of magnetic reversals also support the theory of seafloor spreading, which describes the way in which new oceanic crust is generated at mid-oceanic ridges (see PLATE TECTONICS). As the molten rock cools while moving away on either side of the ridge, the current magnetism is recorded. This eventually produces a striped effect, with each stripe corresponding to a magnetic reversal. Reconstructing geologic history from fossil magnetic data is best accomplished for rocks younger than Triassic. Data from older rocks is less reliable and more difficult to interpret.

M. NAGEL

See also: EARTH, PLANET; ELECTROMAGNETISM; MAGNETISM.

Further reading:
Press, F. and Siever, R. *Understanding Earth*. New York: W. H. Freeman and Company, 1993.
Strahler, A. H. and Strahler, A. N. *Modern Physical Geography*. New York: John Wiley & Sons, 1992.
Vogel, S. *Naked Earth*. New York: Dutton, 1995.

MAGNETISM

Magnetism is the force, related to the electrostatic force, produced by the motion of electric charge

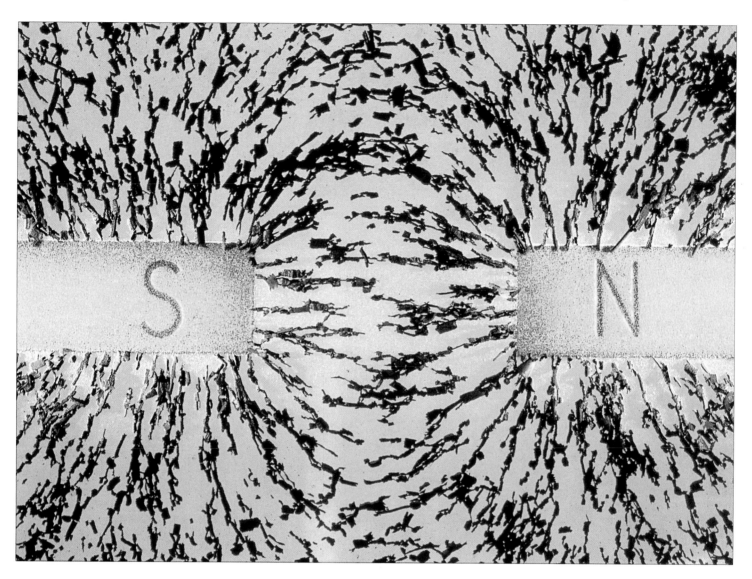

When the unlike poles of two bar magnets are aligned, iron filings show the attractive magnetic field between them.

Magnetism is one aspect of the branch of science called electromagnetism (see ELECTROMAGNETISM). For centuries it was believed that electricity and magnetism were separate forces. It was only as recently as 1905 that German-born physicist Albert Einstein (1879–1955) established in his special theory of relativity (see RELATIVITY) that both are aspects of one common phenomenon.

CORE FACTS

- Magnetic force, like electrostatic force, obeys the inverse square law, decreasing with the square of the distance.
- A magnetic field is produced by electric charge in motion.
- Ferromagnetism is due to the motion of electrons inside the atom.
- Ferromagnetism is characterized by the formation of magnetic domains that can be drawn into alignment by an external magnetic field.
- Iron is the most important ferromagnetic element, although many ferromagnetic alloys do not contain iron.

The history of the study of magnetism

The first known magnets were pieces of mineral that came to be called lodestone. This shiny black mineral, also known as magnetite, is a natural oxide of iron, Fe_3O_4 (see IRON AND STEEL), and the Chinese are believed to have been the first to discover, more than 2000 years ago, that freely rotating pieces of this material always pointed in the same direction, with one end toward the North Star. The Chinese, however, were not adventurous navigators, and it was probably Arab sailors who first realized the great importance of this discovery for long sea voyages, eventually passing it onto Europe.

The first scientific study of magnetism in Europe was reported by Pierre le Pèlerin de Maricourt (known as Peter Peregrinus, "the Pilgrim") in a letter, *Epistola ad Sigerum de Foucaucourt militem de magnete*, written in August 1269. He described the properties of the lodestone and how another magnet could be made by stroking it on an iron needle. He named the end of the magnet that pointed north its north "pole," and the other its south pole, and described how like

CONNECTIONS

- **HEAT** can make the **ATOMS** in a magnet vibrate more freely so they are more disordered; this destroys magnetism.

- **NICKEL AND COBALT** and **IRON** are ferromagnetic **ELEMENTS**.

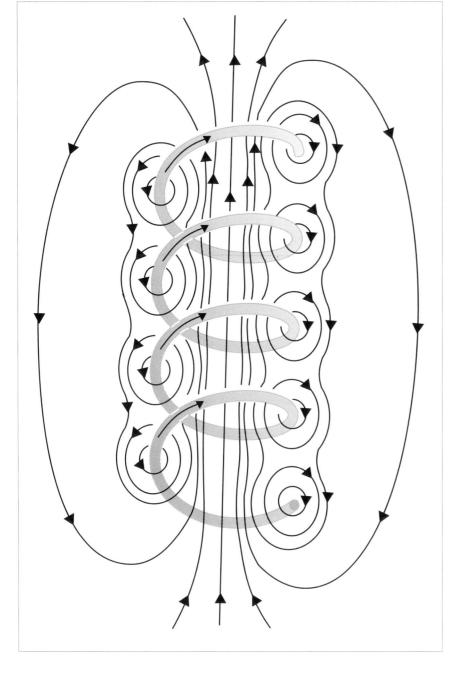

The magnetic field produced by a loosely wound current-carrying solenoid is strong along the axis inside the solenoid but weaker outside it.

THE RIGHT-HAND RULE

To find out the direction of the magnetic lines of force around a straight wire carrying a current, imagine that you are holding the wire in your right hand, with your thumb pointing in the direction of the current. Then the way in which your fingers curl around the wire is the direction of the lines of force.

If the current in the wire exerts a force on the magnet, then the magnet must exert an equal force on the wire (Newton's third law; see FORCES). Imagine a straight wire running between the poles of a horseshoe magnet. Point your right hand, with the thumb upright, in the direction of the current, and then curl your fingers around in the direction of the lines of force—that is, from the the north pole to the south pole. Your thumb now shows the direction of the force on the wire. By convention, electric current is treated as the flow of positive charge, rather than negative. The magnetic force on a moving negative charge is in the direction opposite that on a positive charge moving the same way.

poles repelled and unlike poles attracted one another. He also explained how a magnet was used in direction-finding by floating it on a scrap of wood on a bowl of liquid, and suggested ways of placing it on a pivot.

Later, navigators placed a circle of heavy paper under the pivoted magnet, with radiating lines pointing in 32 directions. To compass means to enclose in a circle, and this diagram was called a compass rose.

The next step

The next major step in the investigation of magnetism came more than three centuries later. English physician William Gilbert (1544–1603) published the results of his detailed experiments in a book entitled *De Magnete* in 1600. He concluded that Earth itself was a giant magnet, and he shaped a lodestone into a sphere to demonstrate this.

Gilbert made a clear distinction between electric (electrostatic) forces and magnetic forces, but early scientists could see that there was a similarity between them: like repelled like and attracted unlike. In England, several 18th-century scientists —John Michell (1724–1793), Henry Cavendish (1731–1810), and Joseph Priestley (1733–1804)— conducted important experiments on both magnets and electrically charged bodies, but the credit for the first detailed report goes to French physicist Charles-Augustin de Coulomb (1736–1806). In 1785, using a delicate torsion balance, Coulomb showed that electrostatic and magnetic forces both obeyed the same law: the force of attraction or repulsion between two magnetic poles or between two electrically charged bodies decreased in proportion to the square of the distance between them.

Further understanding of the relation between electricity and magnetism could not come until the invention of the first electric battery—and so, the production of electric current—by Italian physicist Alessandro Volta (1745–1827) in 1800. In 1820, Danish physicist Hans Ørsted (1777–1851) discovered that current flowing in a wire would deflect a compass needle placed beside it.

News of this discovery quickly spread, and in France André-Marie Ampère (1775–1836) saw a demonstration of the phenomenon given by his compatriot François Arago (1786–1853). Within weeks, Ampère had begun to develop the basis of electromagnetism. He showed that two wires carrying current in the same direction attract each other but repel one another when the currents are in opposite directions. He also showed that a helical coil of wire carrying a current—he called it a solenoid—would behave like a bar magnet and would magnetize a steel needle placed inside it. And he suggested that the deflection of a compass needle could be used to measure the strength of electric current in a wire. This idea soon led to the invention of the galvanometer, the instrument that measures electric current.

In 1820, French scientists Jean-Baptiste Biot (1774–1862) and Félix Savart (1791–1841) made a mathematical calculation of the magnetic field surrounding a long, straight wire carrying a current and showed that the strength of the field is proportional to the current and decreases in proportion to the distance from the wire. This is the Biot-Savart law.

In England, one of the practical demonstrations that Michael Faraday (1791–1867) liked to give in his lectures was to sprinkle iron filings on a sheet of heavy paper and then apply a magnet to them. The filings aligned themselves along curved lines of force between the two poles of the magnet. He could then show how the lines of force were formed in circles around the current-carrying wire. In October 1821 he constructed a device in which a free magnet could rotate about a current-carrying fixed wire, or a wire freely moving about a fixed magnet. It was the first demonstration of the conversion of electrical energy into motion, and it was to lead to the development of the electric motor.

Faraday argued that it should be equally possible to produce an electric current in a wire by means of a magnetic field. In August 1831, almost by accident, he discovered how to do it. He wound two coils around an iron bar, connecting one to a battery to generate a magnetic field and the other to a galvanometer. Nothing happened while the current was flowing, but Faraday noticed that the needle of the galvanometer kicked whenever the current was switched on or off. He immediately deduced that it was the change in the magnetic field that induced the current (see ELECTROMAGNETISM). The effect was discovered at the same time by U.S. physicist Joseph Henry (1797–1878), but he had been too busy teaching to publish his findings.

Meanwhile, in 1825, British scientist William Sturgeon (1783–1850) had made the first electromagnet. He wound a current-carrying wire 18 times around a horseshoe-shaped piece of iron and showed that it would lift more than 20 times its own weight.

Faraday was an experimental scientist and, although he made many of the discoveries that have brought electricity into our modern world, his ideas about the relation between electricity and magnetism remained vague. It was the British physicist James Clerk Maxwell (1831–1879) who followed up Faraday's work. He showed that electric and magnetic fields travel together through space as waves of electromagnetic radiation, with the changing fields mutually sustaining each other (see ELECTROMAGNETIC SPECTRUM).

What is magnetism?

When he discovered that an electric current can produce a magnetic force, Ampère suggested that perhaps a material such as lodestone was magnetic because it contained microscopic loops of current in its structure. He was not completely wrong, although a full understanding of the cause of magnetism had to await 20th-century discoveries in subatomic physics.

Just as an electrostatic charge produces a field around itself and exerts a force on other charges in that field, so a moving charge—such as an electric current—produces a magnetic field. This magnetic

---

MAGNETIC STORMS

Most of the time, the magnetosphere (a magnetic field around Earth), extending about 30,000 miles (50,000 km) into space, acts like a windshield to the solar wind, the stream of ionized particles blowing away from the Sun (see SUN) at about 240 miles per second (400 km/s). If the solar wind gusts, it can upset the electric currents normally created by the interplay of Earth's magnetic field and the charged particles. Powerful new currents suddenly swirl and, with them, magnetic disturbances. Often accompanied by spectacular auroras in the ionosphere, the disturbances interfere with radio, television, and radar, making communications and navigation difficult, and induce currents in electric power grids, sometimes resulting in massive blackouts.

These magnetic storms arise during intense solar flare activity on the Sun's surface and are associated with the 11-year sunspot cycle (see the box on page 700). When sunspots are most common, magnetic storms are most likely to strike Earth, lasting from several hours to 10 days.

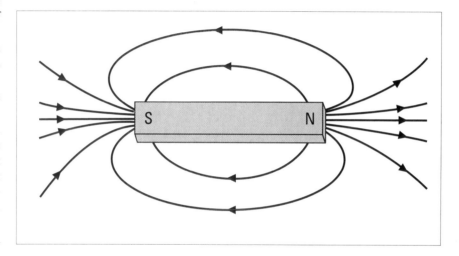

field—represented by the letter B and whose SI unit is the tesla (T)—exerts a force on other moving charges. There is one important difference, however. Electric charges, whether positive or negative, exist independently of each other. A magnet, on the other hand, always has a north pole and a south pole: it is a dipole. If a current loop is placed in a magnetic field, the dipole created by the loop will try to turn to align with the field. This is the basic principle of the elec-

The magnetic field of a bar magnet is strongest at the poles.

STAR DEATH, MAGNETISM, AND PULSARS

About once every 50 years, a massive star explodes in a supernova (see NOVAS AND SUPERNOVAS), throwing off its outer layers. Gravitation then pulls the remaining core inward, collapsing it to a diameter of about 6 to 18 miles (10 to 30 km) and squeezing its matter so compactly that the heart of the star consists almost entirely of neutrons at a density equivalent to that of the atomic nucleus. In shrinking, this neutron star also spins rapidly, as much as 100 times per second. The star's original magnetic field concentrates and spins too. As the magnetic field gets more intense, charged particles above the surface of the neutron star accelerate, releasing electromagnetic radiation like the beams from a lighthouse. The neutron star's radio signals reach astronomers' radio telescopes as rapid, regular pulses, so these flashing neutron stars are called pulsars (see PULSARS). As they broadcast their radio beams, pulsars lose energy and gradually slow down. After about 10 million years, their radio beacons grow too weak to detect.

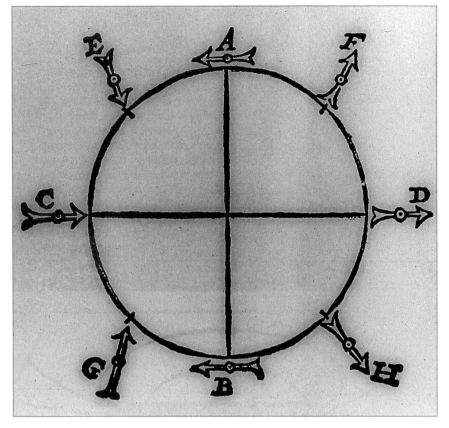

In this 16th-century engraving of Earth's magnetic field (above), line AB is the equator, C is the north magnetic pole, and D is the south magnetic pole. The arrow tips are north-seeking, and arrow directions indicate the way a freely rotating compass would point.

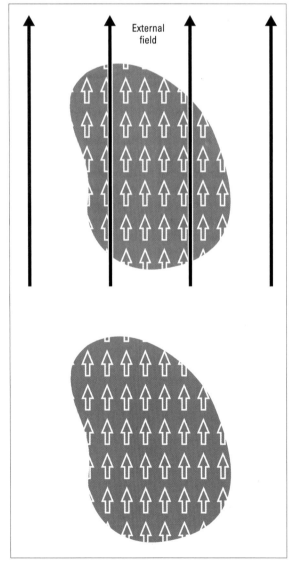

When iron is put into a magnetic field (top), the magnetic moments of electrons line up with the field. When the external field is turned off (bottom), the electron magnetic moments remain aligned.

tric motor. The force that tends to turn the loop is proportional to its area and the current flowing around it. The product of these quantities is called its magnetic dipole moment.

Ferromagnetism

Inside an atom, each electron orbiting the nucleus can be viewed as forming a current loop, just as Ampère suggested, and will therefore produce its own magnetic field. (This is a simplified way of looking at things, and quantum mechanics has made the picture a bit more complicated.) Each electron will have a dipole moment, but usually, inside the atom, most of these moments will cancel each other out.

However, in certain atoms or their ions, the electrons of the outer shell combine their magnetic moments to give the atom itself a strong magnetic moment. Iron, nickel, and cobalt are the best known of the elements that show this effect, which is called ferromagnetism. At normal temperatures, gadolinium and dysprosium are the only other metals with this property. The strength of the magnetism that a material can develop is called its permeability. Nickel's permeability is 40, cobalt's is 55, but that of iron is over 1000. This is why many magnets are made mainly of iron. A number of alloys, including many involving aluminum, nickel, and cobalt (alnico), make even stronger permanent magnets.

But not each piece of iron, for example, is a magnet. This is because, although individual atoms line up to form tiny magnetic regions called domains, these domains are often jumbled up, all pointing in different directions and canceling each other out.

A piece of iron can be magnetized by stroking it with a magnet or by placing it inside a coil carrying a current. This causes the individual domains to line up with one another. In fact, clicking or hissing noises made by the domains aligning themselves can be heard with suitable amplification equipment. This is known as the Barkhausen effect, after its discoverer, German physicist Heinrich Barkhausen (1881–1956). Even Earth's magnetism will magnetize a ferromagnetic material. If an iron rod is aligned with Earth's field and hit with a hammer, the mechanical force is enough to cause the domains to move, and the rod becomes magnetized.

Iron and the other ferromagnetic metals can be obtained with different crystal structures. In some of these, the alignment of the domains does not persist when the magnetic field is removed. Relatively pure iron called "soft" iron, for example, is used in electromagnets and solenoids because the domains will line up easily, but they become disordered just as easily when the current is switched off. In other forms of iron, particularly steel, the alignment of the domains becomes fixed, and these are used to make "permanent" magnets. Even these magnets, however, must be treated with care. They can gradually lose their magnetism if they are struck forcibly, because they will be affected by Earth's magnetic field.

Heat can also destroy a magnet. The atoms gain extra energy and vibrate more freely, so that the domains become disordered. Above a temperature known as the Curie point, all magnetism is lost. The Curie point for iron is 1418°F (770°C); for cobalt it is 2039°F (1115°C); and for nickel, 676°F (358°C).

Paramagnetism and diamagnetism

Even nonferromagnetic materials respond to a magnetic field. Materials in which the molecules or ions have a small magnetic dipole moment are called paramagnetic. Normally they show no magnetism, but if they are placed in a magnetic field, they will tend to align parallel to the field and slightly increase its strength. Ferromagnetic materials with a temperature above their Curie point also behave like paramagnetic materials.

Materials in which the molecules or ions have no dipole moment are called diamagnetic. When they are placed in a magnetic field, a dipole is induced in the molecule or ion, but this will be in the opposite direction to the field and is very weak. Water, many salts, the noble gases, and most organic compounds are examples of diamagnetic materials.

Magnetization and demagnetization

Permanent magnets are made by placing suitable pieces of iron or steel or a magnetic alloy inside the magnetic field of an electric coil. The permanence of the magnetization depends on the exact composition of the magnetic material. Types of steel in which the domains become more rigidly aligned are preferable. Many permanent magnets are made of alnico or other alloys.

"Soft" iron is easily magnetized by an electromagnetic field but loses much of its magnetization when the field is switched off or reversed. This property is very important in solenoids inserted in electric circuits that require rapid switching on and off. The starter of a car, for example, needs a large current to turn it, but once the engine has fired it is no longer needed, and the engine can continue to fire on a much lower current. Turning the ignition key allows a small current to flow through the solenoid: this causes the iron core to become magnetized, and it moves in the magnetic field to close a circuit that feeds high current to the motor. When the key is released, the core is no longer magnetized: it falls back, the circuit is broken, and the motor stops turning.

Inside a solenoid, the strength of the magnetic field of the coil and the magnetic field induced in the iron core are combined to produce a stronger field than that of the solenoid alone: this is why electromagnets can be so powerful. If the current in the electric coil is then reversed, some of the magnetism in the core—but not all—will also be reversed. To destroy all this residual magnetism, the current may have to be reversed, forward and backward, and gradually decreased, many times.

If a graph is plotted, with one axis representing the total magnetic field strength and the other the field strength of the coil, it does not just go up and down in a straight line as the current is reversed. The total field strength lags behind the field strength of

WILLIAM GILBERT AND MAGNETISM

English physician William Gilbert (1544–1603) conducted the first true scientific research on magnets. His findings changed the way people had viewed magnetism for thousands of years.

One of London's most prominent doctors (Queen Elizabeth I was among his patients), Gilbert still found time to study the lodestone in his private laboratory. He published his results in *De magnete* in 1600. In this landmark book, he first dispelled superstitions about lodestones—for example, that they could cure disease or lose their power near goat's blood—and then carefully distinguished their power from the attraction derived from rubbing substances such as amber, which he termed *electrics*. More important, he argued that Earth itself is a gigantic magnet, explained the actions of compass needles, described three ways to make artificial magnets, showed how to improve the power of bar magnets, and proposed that a magnet's power comes from an invisible "orb of virtue" that surrounds it, an idea that, although vague, anticipated the theory of an invisible magnetic field. *De magnete* was a scientific best-seller of the day, inspiring later investigators. Italian scientist Galileo Galilei (1564–1642) was an admirer of Gilbert's experimental methods and painstaking observations.

DISCOVERERS

THE SUN'S MAGNETIC FIELD AND SUNSPOTS

No one fully understands why spots appear on the Sun, even though astronomers have noted them for over 2000 years. Sunspots look dark because they are about 1500 K cooler than the luminous solar surface (photosphere) around them. Their numbers rise and fall through a 11-year cycle, and appear as bipolar pairs; that is one sunspot of each pair has a roughly comparable area covered by north and south magnetic polarities.

Logically, then, sunspots originate from the Sun's powerful magnetic field. The leading theory involves the Sun's differential rotation. (The Sun does not have a rigid suface like Earth and it is believed that regions near the solar equator rotate faster than region near the poles.) This rotation induces the magnetic field to concentrate on either side of the solar equator. In these regions the concentrated field inhibits the movement of atoms, slowing down the convection currents that carry heat from the interior to the surface. Because less heat and light escape, these regions are seen as sunspots, which on average are twice as wide as Earth and last for a week. These dark spots have been detected on other stars as well.

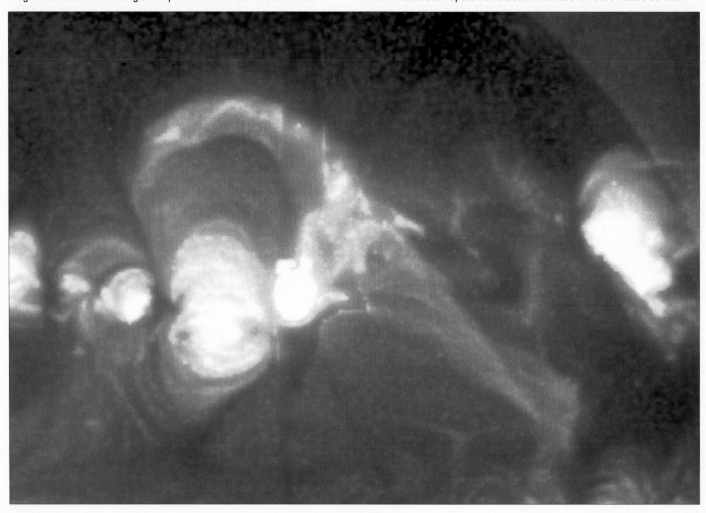

ADIABATIC DEMAGNETIZATION

When atoms or molecules in a substance are aligned by a magnetic field, their random vibration is reduced. This vibration is their thermal energy, and so they will give out heat on magnetization. Researchers have found a way to reach very low temperatures in this way. They surround a paramagnetic salt, such as gadolinium sulfate, with liquid helium and then apply a magnetic field. The heat produced is absorbed by the helium, which is then removed. As dictated by the first law of thermodynamics (see THERMODYNAMICS), when the magnetic field is reduced no heat can flow back into the gadolinium sulfate—this is called an adiabatic process—so its temperature falls below that of the liquid helium. Temperatures as low as 0.0015 K have been reached using this process.

A CLOSER LOOK

the coil, as the domains only gradually become aligned in the reverse direction. The S-curve that is plotted is called the hysteresis curve of the core material (see the diagram on page 702). Hard materials are suitable for permanent magnets because they show a wide, open curve (they retain their magnetism even when there is no current to induce a magnetic response); those used in solenoids show a narrow curve that is close to a straight line (because the magnetic field is more easily switched off and can be reversed with less loss of energy) and are known as soft materials.

Magnetohydrodynamics

Magnetohydrodynamics (MHD) is the merger of two major studies in physics: electromagnetism and hydro-dynamics (fluids and gases in motion; see FLUID

MECHANICS). Movement of charged particles in a fluid induces a magnetic field, which intersects at right angles with a fixed magnetic field, producing an electromotive force (EMF). This mutual interaction is the basis for MHD.

Already used to control plasma (ionized gas; see PLASMA) in thermonuclear fusion research, MHD phenomena will have further applications in advanced technology. If a plasma is forced at high speeds through a magnetic field, it generates a large electric current that can be conveyed through electrodes into an elec-

trical circuit. This MHD generator has several advantages over traditional electrical generators, whose turbines produce current with revolving armatures and contact brushes (see ELECTROMAGNETISM): it replaces the turbine altogether, it can operate at higher temperatures without wearing out parts, and it is highly efficient. With suitable sources—such as nuclear reactors—to heat the plasma, MHD generators are likely to become a central part of commercial power grids.

Research is also under way to put MHD to work as propulsion for ships. Salt water—a good conduc-

MAGLEV

Magnetically levitated (maglev) trains, with speeds of more than 300 mph (480 km/h), may soon be able to compete with air transportation over short distances. They have no frictional wheels running on tracks, instead they float silently along a guideway on a magnetic cushion. Electromagnets inside the train are each attracted by an electromagnetic coil ahead and repelled by a coil behind. The current in the coils is rapidly reversed so that the train can be pulled and pushed forward. At the same time, the repulsive force keeps the train floating inside the guideway. Prototype maglev trains are already operating in Germany and Japan.

SCIENCE AND SOCIETY

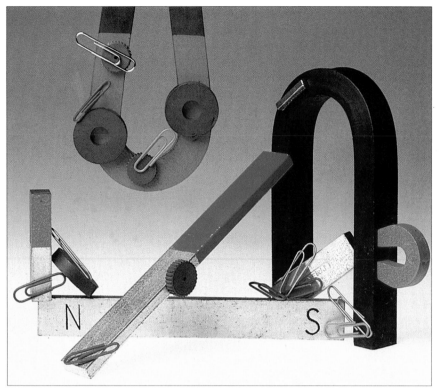

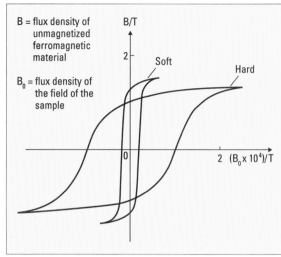

B = flux density of unmagnetized ferromagnetic material

B_0 = flux density of the field of the sample

Hard magnetic materials, such as permanent magnets, have a wide, open hysteresis curve. Soft magnetic materials, such as those in solenoids, have a narrow curve.

Magnets come in many shapes and sizes.

tor of electricity as it contains ions—is passed between superconducting magnets, which accelerate the water, providing thrust. This potentially powerful and quiet propulsion system became famous through Tom Clancy's technothriller novel *The Hunt for Red October* (published in 1984) in which an MHD-powered Soviet missile submarine is undetected by its pursuers. The idea is not purely fiction, however. In 1992, a Japanese company tested MHD propulsion. Its *Yamato*, a tiny craft, managed about 6 mph (10 km/h), but some scientists predict speeds of up to 100 mph (160 km/h) in the 21st century.

MHD already has an immense natural influence on Earth. Its effects influence the movement of charged particles in the Van Allen radiation belts (see VAN ALLEN BELTS) and allow the dynamo at the planet's center to create its magnetic field.

MAGNETOTACTIC BACTERIA

Magnetic personality has an important meaning for the survival of some bacteria. Earth's magnetic field is a map to their food. Grains of magnetite in their cells let them sense the direction of magnetic field lines, an ability called magnetotaxis. Living in lakes and swamps, the bacteria swim in the field's direction, which dips toward the poles, until they come across the sediments that are their normal feeding grounds. Bacteria in the Northern and Southern Hemispheres have evolved to follow the field lines in different directions because of the different polarities of Earth's field. If northern bacteria are placed in a southern environment, sure enough, most swim in the wrong direction and starve, but not all. Enough wrongheaded ones, probably mutations, will inadvertently swim in the right direction and so ensure that a population of bacteria survives.

Another species, known as GS-15, is believed to be responsible for making lodestones. Deep under ground, they get the oxygen they need by converting ferric oxide (Fe_2O_3) to magnetite (Fe_3O_4). During the last billion years, they have built up, layer upon layer, throughout the world and under the seas.

Other uses of magnetic fields

Magnetism is used to obtain high-resolution pictures of the body's internal soft tissue. In a technique called magnetic resonance imaging (MRI; see RESONANCE), the patient is exposed to a very strong magnetic field, of the order of a few tesla. Hydrogen nuclei in the body of the patient act like tiny magnets and the applied magnetic field causes them to line up with the field, but also resists the movement. The result is a precession, or "wobble," which has a characteristic frequency. If an electromagnetic field (radio wave) oscillating at the precession frequency is applied to the particle, the particle will absorb or emit energy, often changing its orientation. A suitably tuned radio detector can detect the absorbtion or emission and processes the information into a computer image. Soft tissues cannot be seen well using other imaging methods, and, unlike X rays, magnetic fields do not have any damaging physiological effects.

Magnetic forces also allow us to hear music. A stereo speaker produces sound waves when its stiff, lightweight paper cone vibrates. This vibration is caused by a varying magnetic field applied to the cone. When a coil of wire is wrapped around a hollow cylinder at the end of the cone, this cylinder slides back and forth inside a narrow cylindrical slot in the magnet. This produces a magnetic field. The coil is connected to the stereo's amplifier, which provides a signal in the form of a time-varying current. The magnetic field exerts a force on the coil, and this varies according to the magnitude and direction of the current in the coil.

R. SMITH

See also: ELECTRICITY; ELECTROMAGNETIC SPECTRUM; ELECTROMAGNETISM; ELECTRONS AND POSITRONS; ELECTROSTATICS; MAGNETIC POLES.

Further reading:
Verschuur, G. L. *Hidden Attraction: The History and Mystery of Magnetism*. New York: Oxford University Press, 1993.

MANGANESE

The name *manganese* comes from a misunderstanding followed by an accident. The element occurs naturally as an oxide, pyrolusite, which is found as shiny black crystals, not unlike those of magnetite (magnetic iron oxide, Fe_3O_4). Magnetite comes from the ancient Greek district of Magnesia, and, sometime during the 16th century, the word *magnesia* became converted in Italian to manganese. This may even be a play on words. For two centuries after 1504, the Italian district of Naples was a Spanish kingdom, and the Neapolitan dialect still contains many Spanish words. In Spanish, the word *mangano* means "trickster," and the fact that pyrolusite was not magnetic iron oxide may be reflected in this name.

In 1774, Swedish chemist Carl Wilhelm Scheele (1742–1786) was the first to recognize manganese as a separate element, and in the same year it was isolated by his associate Johan Gottlieb Gahn (1745–1818).

Manganese is the 12th most abundant element in Earth's crust, making up about 0.1 percent of its mass. Like most metals, manganese—atomic number 25—belongs in the transition series between groups 2 and 13 (IIA and IIIA) of the periodic table. As a member of group 7 (VIIB), manganese lies between chromium (24) and iron (26). The only other members of group 7 are the unstable radioisotope technectium (43), the rare metal rhenium (75), and the artificial element unnilseptium (107).

Physical properties

At room temperature, manganese is a gray-white, hard, brittle metal, with a high melting point of 2280°F (1247°C). It has no practical uses itself but is very important as an ingredient of steel.

Chemical properties

Manganese is somewhat more reactive than iron and chromium, its neighbors in the periodic table, and its compounds are more diverse, owing in part to the

Swedish chemist Carl Wilhelm Scheele, about age 25. In 1774, Scheele was the first to recognize that manganese was a separate element.

fact that manganese possesses a wide variety of oxidation states—from -3 to +7. For example, in manganese oxide (MnO) the oxidation number of manganese is +2. In manganese dioxide (MnO_2) it is +4. In potassium manganate (K_2MnO_4) it is +6. In the permanganates (salts containing the anion MnO_4^-), such as potassium permanganate, $KMnO_4$, it is +7.

Because the deep purple permanganates are strong oxidizers, they have found various important uses both in the laboratory and in industry. For example, in industry, permanganates are used in the production of such commercial products as saccharin and benzoic acid. Permanganates are also used in some water purification systems as disinfectants and, unlike chlorine, do not impart an unpleasant taste to the water. In addition, permanganates tend to produce particles of MnO_2 in water. These act as coagulants that absorb various particulate impurities in the water, which can then be mechanically removed. In the laboratory, permanganates are used in a variety of analytical applications.

CORE FACTS

- Manganese is an electropositive transistion element of group 7 (VIIB) of the periodic table.
- At room temperature, manganese is a gray-white, hard, and brittle metal that has a high melting point of 2280°F (1247°C).
- Manganese possesses a wide variety of oxidation states—from -3 to +7.
- Manganese is the twelfth most abundant element in Earth's crust, comprising 0.1 percent of its mass.
- The primary use of manganese is as a deoxidizing and desulfurizing agent in the manufacture of steel, but it is also used to make other alloys, ceramics, dyes, fertilizers, glass, paints, and batteries.

Symbol: Mn
Atomic number: 25
Atomic mass: 54.9380
Isotopes:
   55 (100 percent)
Electronic shell structure: [Ar] $3d^5 4s^2$

CONNECTIONS

- Manganese is a reactive **METAL** that readily combines with **OXYGEN** and the **HALOGENS** when heated and reacts slowly with **WATER**.

- **DYES** and certain **PAINTS AND PIGMENTS** contain manganese compounds.

MANGANESE

MINING MANGANESE FROM THE OCEAN FLOOR

Tremendous reserves of manganese lie on the seafloor, especially on the bottoms of the Indian and Pacific Oceans. These resources are in the form of round rocks called nodules, which consist largely of manganese, iron, cobalt, copper, and nickel oxides. Although 24 percent of the nodules is manganese, the potential value lies more in their nickel and copper content, because these metals are relatively rare on the surface of Earth. Engineers have cast an eager eye on the nodules as a valuable source of minerals. However, no economically feasible method of mining the nodules has been developed, and efforts continue to find a profitable way of bringing the nodules up from the bottom of the sea.

Manganese nodules on the seafloor, 16,000 ft (4900 m) below the surface of the Pacific Ocean.

SCIENCE AND SOCIETY

Potassium manganate (K_2MnO_4) is deep green, but even weak acids, such as carbon dioxide (CO_2) will convert it into the purple permanganate and the dioxide:

$$3K_2MnO_4 + 2CO_2 \rightarrow 2K_2CO_3 + 2KMnO_4 + MnO_2$$

In solution, the color changes rapidly from green to pink, which has caused the manganate to be named the "mineral chameleon."

Large pieces of manganese are relatively stable chemically, although oxidation will occur on the surface. However, manganese is much more reactive if finely ground. For example, powdered manganese—particularly if it is contaminated with small quantities of carbon—will react with water to form a hydroxide and to liberate hydrogen gas. It will also react with weak acids, such as dilute HCl, to form salts: in this case manganese chloride ($MnCl_2$).

Occurrence and manufacture

Although manganese ores occur in the United States—chiefly in Chamberlain, South Dakota; Cuyuna, Minnesota; Aroostook County, Maine; Artillery Peak, Arizona; and Butte, Montana—most are of a low-grade variety, which makes the United States a manganese "have not" nation. This means that more than 90 percent of the manganese ore used by the United States is imported from abroad, mainly from South Africa, France, Australia, and Brazil.

The major technique employed to produce manganese metal involves the electrolysis of manganese sulfate solutions. In this process, manganese ores—mainly those that contain MnO_2—are first roasted in ovens to produce manganese oxide. The oxide is then treated with sulfuric acid, which converts the oxide into manganese sulfate. Impurities such as iron, arsenic, tin, lead, and cobalt are then removed. The sulfate is dissolved in water and the resulting solution is led into the cathode chamber of an electrolytic cell. An electric current then causes manganese to collect at the cathode. The manganese, which is 99.9 percent pure, is collected from the cathode.

Uses of manganese

The primary use of manganese is in the manufacture of steel (see IRON AND STEEL). Manganese performs two basic functions in steel production: it increases the hardness of the steel and it reacts with and removes impurities such as sulfur and oxygen. Such impurities make the steel brittle.

Manganese in the form of its ore, the dioxide MnO_2, formerly used in the manufacture of chlorine, was first used in glass as far back as the time of ancient Egypt. It is still used for that purpose. The compound gives the glass a pinkish tinge; it can be used to bleach glass—the pink counteracts the green tinge caused by iron impurities. Manganese dioxide is also used in the manufacture of carbon-zinc dry batteries (see BATTERIES), where it inhibits the unwanted formation of hydrogen gas at the carbon electrode. Manganese dioxide, mixed in various proportions with materials used to make bricks, can impart hues of red, brown, and gray.

Finally, small amounts of manganese are added to metal alloys, other than steel, to improve their properties. For example, manganese is added to aluminum alloys to improve their resistance to corrosion and to make them harder. Manganese is also mixed with such metals as nickel, zinc, and copper to produce alloys for special applications (see ALLOYS).

C. PROUJAN

See also: ALLOYS; ANALYTICAL CHEMISTRY; BATTERIES; IRON AND STEEL; TRANSITION METALS.

Further reading:

Atkins, P. W. *The Periodic Kingdom.* New York: Basic Books, 1995.
Barrett, J. *Understanding Inorganic Chemistry*. New York: Ellis Harwood, 1991.

MAPS AND MAPPING

Mapping is the process of making maps, graphic representations of parts or all of Earth's surface

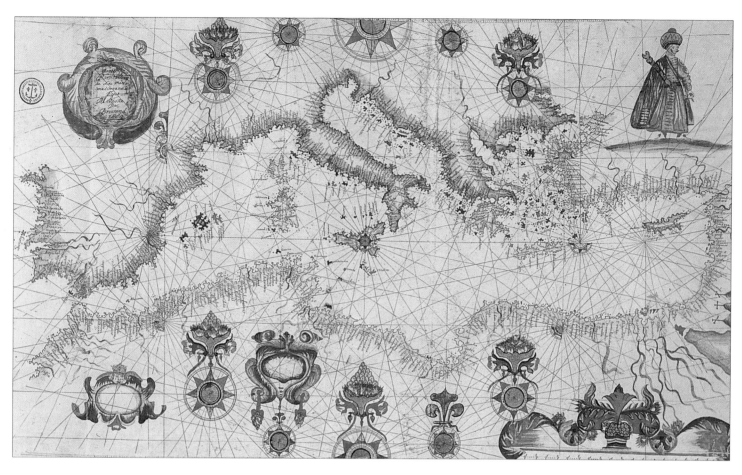

This colorful map of the Mediterranean was produced in 1664.

Maps present information of almost any kind on a geographical layout based on some portion of Earth's surface. Except on a globe, the whole surface of Earth cannot be displayed without being distorted. Most maps present a limited area for a specific purpose. World maps require selection of a projection that will best represent the portions of the world important to the theme of the map. A projection represents the curved surface of Earth on a flat map. There are many projection styles.

Cartographers are mapmakers. Mapmaking (mapping) has been practiced since ancient times in all parts of the world and using a variety of mediums, from sticks and shells to silk. Paper maps are the modern standard, but computer screen images are widely used. The first element in mapping is the acquisition of data about the surface to be mapped. This is done by a survey team or by using satellite imagery. The historical development of topographical mapmaking has largely been controlled by the technical advances in instrumentation for accurately measuring the frame of reference required: compass directions, levels to the horizon, and height above sea level, for example.

The cartographer also needs to select the scale for the map. This scale defines how much physical space is represented by a unit of space on the map. Scaling down can be simplified by using symbols to represent information. Some map symbols are widely used (see the box on page 710).

Map subjects

Specific data relevant to the theme of the map must also be acquired. Theme maps are almost limitless. Common maps include political and travel maps, street maps, weather maps and other scientific maps such as geological maps, natural resources maps, and topographical maps. Transportation maps, such as railroad, under ground and above ground, and maps of waterways are common. Maps of large bodies of

---

CORE FACTS

■ Almost any kind of information can be presented on a geographical layout of some portion of Earth's surface; the layout is called a map.

■ Maps have been used in many parts of the world since ancient times.

■ Mapmaking has improved with the use of instruments for surveying to collect data, satellite imagery for base data, and computers to store and manipulate data.

■ Some distortion is unavoidable in whatever way curved areas of Earth are projected onto a flat map.

■ Map symbols and scales are used to present information in a concise format.

CONNECTIONS

● Maps can show the **CLIMATES** of different countries and the **WEATHER** in a place at a particular time.

● **REMOTE SENSING**, using **SATELLITES** in **SPACE**, enables different **LANDFORMS** to be recorded on maps.

ATLASES OF MAPS

Mercator introduced the term *atlas* to mean a collection of maps. He chose the word from Greek mythology for Atlas, the Titan who was forced to support the heavens on his shoulders as a punishment for warring against Zeus. A picture of Atlas supporting the world was often a feature of early map collections. Today, we use the term *atlas* to refer to a thematic collection of charts or maps that are systematically assembled in book form or on computer software. World atlases and road atlases are most popular, but there are many other themes used. These include regional areas and historical places, as well as political, social, cultural, and biological issues. There are atlases of the oceans and of the heavens. Since 1968, there have been more than 11,000 books cataloged with the name *atlas* in their titles at the United States Library of Congress.

HISTORY OF SCIENCE

water presenting navigational information are called nautical charts and are generally very detailed. In the wider definition, there are maps of the sky and planets.

Early maps and mapping

Maps have been made since ancient times by people all over the world, including those who have no written language. Isolated populations used whatever they could. Before European influence, people of the Marshall Islands in the central Pacific were using stick charts. Sticks were tied together in patterns with palm fibers to represent prevailing winds and waves. Islands were denoted by shells and corals.

The oldest known maps come from the Middle East. One is a small clay tablet map that shows

This map depicts the world as it was understood around the year 168 B.C.E.

streams, settlements, and hills. North, east, and west are identified. The map has been dated at about 2500 B.C.E. It was found in what is now northern Iraq; it was made to record land holdings in what at the time was the city-state of Akkad in Mesopotamia. Cadastral mapping, the mapping of boundaries as for titles and taxes, dates back to that tiny Mesopotamian tablet.

Silk maps from the second century B.C.E. have been found in the Hunan province of China. Map features were painted on silk that had a faint impression of a grid woven into it. Ancient Greeks were more philosophers than cartographers, but their curiosity did lead to speculation about Earth's size and shape. Aristotle (384–322 B.C.E.) promoted the concept of a spherical Earth. About the time the Chinese were making silk maps, Greek astronomer and geographer Eratosthenes (c.276–c.194 B.C.E.) used a geometrical approach to calculate the circumference of Earth and came remarkably close to the measurement modern scientists calculate today.

Eratosthenes drew parallel east-west lines on a map of the world he had prepared. Fifty years later, Hipparchus, a Greek astronomer, added north-south lines at right angles to the east-west parallels. By the second century B.C.E., Claudius Ptolemaeus (Ptolemy) was using the terms *latitude* and *longitude*. Latitude is an angular distance in degrees north or south of the equator, where the latitude is 0 degrees. Longitude is the distance in degrees east or west of the prime meridian, at 0 degrees. (Meridians are imaginary lines joining the North and South Poles at right angles to the equator. They all meet at the geographic poles. The length of the meridians equals half the world's circumference. Meridians are also assigned degrees of longitude, and Greenwich, England, is 0 degrees.) Longitude is measured by the angle between the plane of the 0-degree meridian and that of the meridian through the point concerned.

Ptolemy wrote a great deal about mapping and the problems of projecting features from a spherical surface onto a flat map. A world map and 26 regional maps were included in Ptolemy's book *Geographia*.

Geographic revival

There was not much progress in mapmaking during the Dark Ages in Europe. It was not until 1295, when Marco Polo returned from his famous journey to China, and when the Renaissance was beginning, that people became interested in maps again. At that time, fairly good charts of familiar waterways and harbors were available for the well-traveled routes, but land maps and world maps were still as Ptolemy had left them, or they were based only on legend. Following the discoveries by Christopher Columbus, the 16th century saw a surge of interest in exploration and cartography.

The development of the printing press and copperplate printing made it easier to produce a large quantity of maps more quickly. Map publishing houses like that of Gerardus Mercator (1512–1594), a Flemish geographer and cartographer, flourished. By the late 17th century, maps improved because better data was available through the use of a surveying technique known as triangulation. In triangulation, a base line is accurately measured between two points. The angle from each end to a third distant point is measured, and then this information is used to calculate the length of the two remaining sides of the triangle.

In 1669, the French Royal Academy of Sciences commissioned Jacques Cassini (1677–1756) to survey and map all of France by triangulation. He died before the task was complete, and his son César-François Cassini de Thury (1714–1784) continued the surveying work and also began construction of a topographical map of France. In cooperation with the British Royal Society (formed to promote scientific research), the counterpart to the French Academy, a map of the English Channel was completed by triangulation in 1790. An instrument called a theodolite was used in the mapping. This is a surveying instrument that measures both horizontal and vertical angles. From about 1720, theodolites were fitted with telescopes, making them useful for long-distance measurements. The telescope was small, tripod-mounted, and free to move in horizontal and vertical planes. The 19th century saw the first complete and reliable, instrumentally based, geological surveys and maps of whole countries.

Great changes in mapping resulted from the two world wars in the 20th century. Aerial photography first began during World War I. Photogrammetry, the taking of measurements from photographs, increased the quality of mapping, particularly in difficult areas. Stereophotography from the air provided vertical images, taken from points of view a little way apart, that could be combined to provide a view in three dimensions, allowing the measurement of relative height and the detailed mapping of topography. During World War II, demand for military maps, and for maps to inform the general public, led to more refinements in mapping. With the advent of satellites, space imagery was introduced. Advances in deep-sea echo sounders (sonar; see MARINE EXPLORATION) have made complete ocean floor mapping possible. The continent of Antarctica has been mapped using aerial photography, sonic probes, and gravity measurements. Most recently, airborne radio-echo sounding and satellite measurements have mapped a 3860-sq-mile (10,000-km^2) freshwater lake that lies some 2.5 miles (4 km) below the surface of the Antarctic ice sheet. Digital imagery has computerized modern mapping.

Information sources for mapping

Every map requires base data on the geographic features, boundaries, and area for the portion of Earth's surface that will make up the background of the map. For centuries, base data has been collected by surveying (see SURVEYING). Since World War I, photogrammetry has supplemented land surveys, and since World War II and the widespread use of satellites, space imaging has greatly expanded base data sources for mapping.

Local boundary and road mapping is done by land survey teams, sometimes using instruments that were developed in the 17th and 18th centuries. The three basic measurements are distance, height, and angle. Short distances are measured with a steel tape, similar to the chain that had been used. Modern electronic distance-measuring instruments are used to measure long distances. A combination of a surveyors level and surveyors rod are still used to determine height. Survey measurements of height and distance produce information on the physical configuration of the surface (called relief) that is needed to produce topographical maps. Except for local maps and field-checking for accuracy on larger maps, photogrammetric methods are used to produce modern topographical maps. In land surveying, angular measurements are made with a transit, a theodolite that has been modified so that the telescope can be rotated 180 degrees about a horizontal axis. Triangulation is still used to determine area.

Base data for mapping is stored on computer software known as geographic information systems (GIS). This allows the user to manipulate map data

Flemish cartographer Gerardus Mercator opened a successful map publishing house toward the end of the 16th century.

SATELLITES AND MAPS

The first communication satellite, ECHO 1, launched from Cape Canaveral on August 12, 1960, was an aluminum-coated plastic sphere, which served as a reflector for radio signals. These provided useful data to allow geodetists (scientists who measure Earth) to plot geographical locations more accurately.

The modern era of space imagery was launched with Earth Resources Technology Satellite, ERTS 1, in 1972. That project was renamed LANDSAT in 1975. A succession of LANDSAT satellites has followed. Older LANDSAT satellites carry the Multispectral Scanning System, MSS. Newer ones carry an improved version of the system. Both the old and new multispectral systems can image strips 115 miles (185 km) wide, but the new version provides clearer images. France, Japan, Russia, and India are among the countries that have also launched satellite imaging systems.

Geodetists have improved the accuracy of geographical measurements significantly in recent years using the satellite-based Global Positioning System (GPS). This system provides a means of fixing a position on Earth's surface by using signals from a number of satellites orbiting in space. Establishing an exact location or a particular reference point is useful to navigators of ships and airplanes as well as to land surveyors.

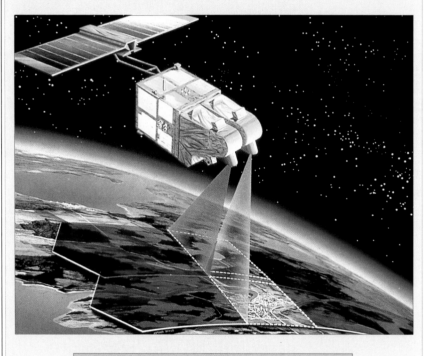

SCIENCE AND SOCIETY

for any theme desired. Base data collected from land surveys and satellite digital imagery is used in computer applications. The governments of Canada, Australia, Britain, and the United States have been working since 1988 on a worldwide GIS database to produce maps of every point on Earth on a Digital Chart of the World (DCW). The Defense Mapping Agency is the source for information on DCW in the United States.

The U.S. Geological Survey's Earth Science Information Center publishes a weekly, called *Sources of Digital Spatial Data*, that describes GIS available from government agencies and the private sector. This modern proliferation of GIS can be compared to the Age of Discovery in Europe when Mercator was producing large quantities of maps.

Space imagery for maps

Solar radiation is reflected in varying amounts from the surface of Earth; the amount depends on the character of the surface. Energy that is not reflected is absorbed, warming up the surface. Night and day, heat is emitted as infrared radiation.

This reflected or emitted radiation can be measured by passive and active space imagery. Passive systems measure the amount of reflected light or emitted infrared radiation. Infrared imaging is called thermatic mapping, and it uses film that is sensitive to the infrared range. Digital imagery from thermatic mapping is used to make geological survey maps. Some satellites are equipped with multispectral scanners that simultaneously record image data from a particular scene in several different wave bands, across the whole range of passive radiation. From these, a whole range of false-color images can be obtained. Such images can be used to show, for example, differences in vegetation cover and soil type.

Active space imagery systems use radar. Pulses of radar (radio) energy are directed at Earth's surface. The image is produced from the varying percentages of the energy reflected from the surface back to the scanner. Radar imaging has the advantage over passive systems of being able to penetrate cloud cover and heavy foliage. It can also penetrate the surface of sandy, arid areas. In fact, some areas of Earth's surface were not really mapped until the development of space imagery. The Amazon River system, for example, was surveyed by German naturalist Alexander von Humboldt (1769–1859) around 1780, but the area beyond it, which makes up almost 60 percent of Brazil, was considered unmappable because it was impenetrable by land (due to the density of the vegetation). With the development of side-looking airborne radar (SLAR) in 1971, the technology for mapping the Amazon was available. In 1981, the Space Shuttle Imaging Radar-A system was used to map the Sahara, revealing a network of ancient river valleys under the sand.

Space imagery evolved as satellite systems were developed using the technology of the space race. Military surveillance, weather, and communication satellites were all developed in the United States in 1960. More recently, geodetists (scientists concerned with determination of the size and shape of Earth) have improved geographical measurements by using the satellite-based Global Positioning System (GPS; see the box at left).

Map projections

Ptolemy in the second century B.C.E. wrote of the need for "adjustments" when making a flat map of a spherical world. The term *adjustments* refers to unavoidable distortions that occur when projecting a curved surface onto a planar one. Over the years, there have been more than 200 different projection techniques proposed, but less than 15 are common. The classic projection for over 400 years has been the one used for world maps that was devised by Gerardus Mercator.

The Mercator projection is described as conformal. The shape of any small area on a conformal map is true to its shape on a globe, but Mercator map areas, though conformal, get relatively larger going toward the poles. This gives a greatly exaggerated impression of the size of countries such as Greenland.

Each mapmaker has to decide what feature is most important to preserve when making a projection. Mercator designed his maps for navigators, and a straight line drawn between any two points on his maps shows the constant-course line, called the rhumb line. He took liberties with the meridians and parallels.

A Mercator projection is rectangular. All the meridians are the same length and perpendicular to the parallels. The parallels are all the same length but longer than the meridians. Parallels on a globe are concentric circles that get smaller in each direction from the equator. Meridians are the same length on a globe, but they are not parallel.

Another type of projection used for world maps is described as equal-area or homolographic. Any area on the globe is represented on the map to the same scale (see the diagram below). The central meridian is straight and perpendicular to the equator. The rest of the meridians are increasingly curved toward the

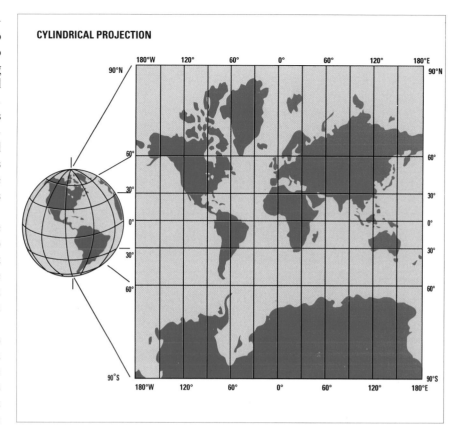

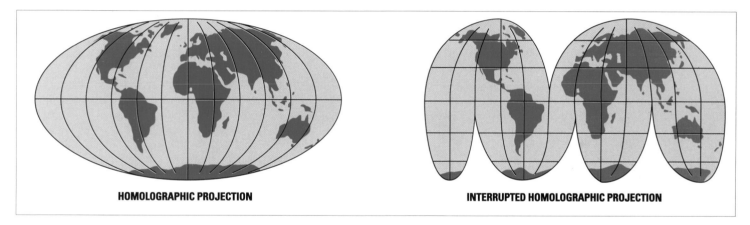

HOMOLOGRAPHIC PROJECTION

INTERRUPTED HOMOLOGRAPHIC PROJECTION

outer edges of the map. All the parallels are straight and decrease in length and separation toward the poles. The distortion of shapes and distances away from the map center is a distinct disadvantage. Some versions of this projection are interrupted, which means strategic cuts are made along meridians to spread the map out, reducing distortion.

There are different techniques used to design a projection. Mathematical formulas and empirical data are both used. There are also three geometrical approaches to mapmaking—cylindrical, conic, and azimuthal. Azimuthal projections are made by projecting latitude and longitude onto a flat surface that touches the globe at one point and is tangential to it (see the diagrams at right). Latitude and longitude are projected onto the surface as they are seen from above the flat surface looking toward the center of Earth. Azimuthal projections are useful for showing polar projections and large continental masses from the equator, but they lead to some distortion at the circumference.

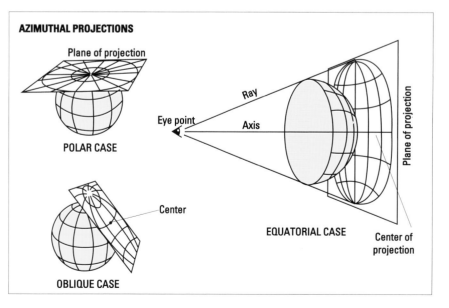

The world can be mapped in a variety of different ways; for example, by cylindrical projections, homolographic projections, and azimuthal projections.

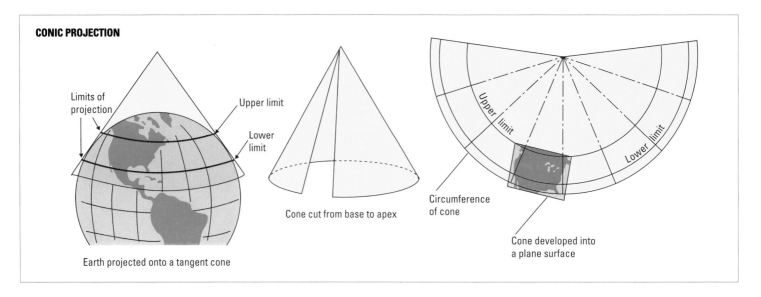

CONIC PROJECTION

Limits of projection

Upper limit

Lower limit

Earth projected onto a tangent cone

Cone cut from base to apex

Upper limit

Lower limit

Circumference of cone

Cone developed into a plane surface

Conic projections are useful for making a map of a country or a part of a continent.

The cylindrical projection makes a rectangular map when the cylinder is cut along a meridian and flattened (see the diagram on page 709). It is produced by projecting the meridians and parallels on a globe onto a cylinder tangent to its surface along a great circle, usually the equator. A Mercator projection is not exactly a cylindrical projection. It is an empirical modification of a cylindrical projection, because a literal transfer of the parallels on a cylinder tangent at the equator would make the parallels closer toward the poles. Mercator changed the scale of his maps with latitude, so the parallels are increasingly farther apart. He did this to maintain conformality. Thus the stretching of parallels of latitude compensated for the east-west stretching. Although shape relationships are maintained, size relationships are not.

Conic projections are drawn onto a cone that is set over the globe like a lamp shade (see the diagrams above). The apex of the cone is above a geographic pole. The cone is tangential (as if it touches the globe)

along a circle, which is called the standard parallel for that projection. Other parallels are drawn as arcs above and below the standard parallel.

Meridians are drawn as straight lines radiating from the apex of the cone. The cone is cut along a meridian and flattened. Only the region of Earth projected close to the standard parallel is useful. Conic projections are common in maps of a country or part of a continent, but they are not useful for world maps.

In theory, no map can have the correct area and be conformal, but it can have a blend of the two properties. In 1988, a compromise projection for a world map developed by Arthur H. Robinson was endorsed by the National Geographic Society of the United States, and in 1989 the American Cartographic Association used one as their official world map.

Scales

A map scale is the mathematical relationship between a specific distance on a map to the distance between the same two points on Earth's surface. Scale on a map may be expressed as a representative fraction (RF) or as a ratio of the distances measured in the same units. Scale is also expressed as a graphic or bar chart comparing a unit on the map in inches or centimeters to the same distance measured on Earth's surface in miles or kilometers. For example, if one inch on the map equals one mile on Earth's surface, it could be expressed $1" = 1$ mile, or $1/63,360$, or $1:63,360$, since there are $63,360$ inches in a mile.

Map scales are often described as large or small. The terms *large* and *small* refer to the relative decimal value of the ratio or RF. The decimal value of $1:10,000$ is much larger than the value of $1:50,000,000$. The scale $1:10,000$ is a large scale that would be useful for mapping an area about the size of the center of a city on a map of only that city. The scale $1:50,000,000$ is a small scale that would be appropriate for mapping the Pacific Ocean from Asia to North America in a standard atlas. Any area covered by a small-scale map will show the unavoidable distortion of the projection selected for the map. Most small-scale maps are prepared to present a gen-

USE OF SYMBOLS IN MAPPING

Maps are graphic representations of a portion of Earth's surface designed to present concise information. Some symbols are often used, but there are no universally accepted ones for all the different maps published. The symbols used for a map are explained in the legend or key.

If the map is part of an atlas, the legend is often contained in the introduction to the collection. Map symbols are intended to communicate information at a glance. Some symbolism, such as the background colors of green for lowlands, brown for elevated land, blue for water, and white for ice and snow, are so common as to sometimes be assumed obvious, and therefore they are not included in the map legend.

Quantitative data, such as the dimensions or population of an urban area, is usually represented by the size of a shaded area at a location on the map. Variations of line style indicate different boundaries. Generally, the broader the line, the greater the significance of the boundary. The same system is used for roadways. Color coding is often combined with shape and form for map symbols. Topographic maps present a lot of symbolic information. City transit maps generally have no scale, no grid, and seldom any more symbols than color coding to schematically indicate the routes of different transit lines.

eral theme, such as city subway maps, so the distortion of scale, and therfore actual distances, on the map is not a problem for the user. A scale given on a world map of 1:100,000,000 does not imply great accuracy, and some mapmakers modify the scale by adding the word *approximate*. Where a linear scale on a map is reliable, the scale is said to be true. The equator on a cylindrical projection and the standard parallel on a conic projection would be true linear scales. Conic projections are considered to produce good approximations of shape, area, and scale for maps of a large country or a continent, but they would not be useful for global maps. The scale on a Mercator projection is sometimes graphically indicated as varying with latitude in the legend or key printed on the map.

Grid reference

A grid is a systematic network of horizontal and vertical lines, usually intersecting at right angles and forming squares, that is superimposed on a map to help people find certain places. The Cartesian planar coordinate system is a grid. It is made from equally spaced lines that are drawn parallel to each of the axes. These lines are labeled with letters or numbers. A point of origin is selected on the plane where two perpendicular axes intersect. The points where these lines intersect can be viewed as corners of squares. The reference of the certain point is a combination of letters and numbers that gives the bottom left-hand corner of the square containing the point.

It is not possible to use a Cartesian planar coordinate system on Earth, since Earth is spherical and the surface of Earth curves away in every direction from any point. Although a network of lines similar to the planar coordinate system was devised by the ancient Greeks, by the time Ptolemy was writing his *Geographia*, north-south meridians of longitude and east-west parallels of latitude were used to systematically locate points on Earth. The spherical coordinate system used on Earth's surface is more often called a graticule than a grid.

The equator is the key parallel, for all the other parallels are concentric circles that get smaller toward the poles. Since before Ptolemy, the parallels north of the equator have been called north latitude, and those south of the equator south latitude. The equator is at 0 degrees latitude, and the poles are at 90 degrees. There has not been such agreement through the years on the prime meridian. Longitude lines go from pole to pole, and they are all the same length. The selection of a prime meridian was a political contest. There was no agreement among nations until 1884, when the meridian that passes through the Royal Greenwich Observatory, England, was selected as zero longitude, although it was not used by all countries until after World War I.

Thematic maps

Maps designed to present specific information beyond boundaries, place names, and roadways are called theme or thematic maps. There is no limit

TOPOGRAPHIC MAPS

Topography, the nature of land surface features such as relief, drainage patterns, and other natural features, is shown on thematic maps. Significant human-made features are also sometimes shown on them. Relief, the variation of elevation and slope of the land, is indicated on a topographical map by contour lines that connect points of equal elevation. The closer the spacing, the steeper the slope. Relief is shown on some maps by plastic shading, which looks like an obliquely illuminated view from above, with light and dark areas giving a three-dimensional appearance. A rugged region is described as having high relief and flat country as having low relief.

Photogrammetric methods are used to collect information for topographic maps. The LANDSATs and similar satellite multispectral scanning systems collect information that is fed to computers to generate stereoscopic images, which can be used to produce a three-dimensional effect in topographical maps. Photo images of relief features frequently replace plastic shading artwork. Before space imagery was available, land surveys were the source of information. Now they are used only to field-check data.

Small-scale maps of large landmasses and global maps sometimes include plastic shading and color coding to show topographic information in the background. A topographic profile may be included with a topographic map. This profile is a cross-sectional view drawn to an indicated, but exaggerated, scale.

Topographic maps of the seafloor have been produced by the advanced technology developed in World War II, and these show midocean ridges and other relief features of the ocean floor not known before.

A CLOSER LOOK

to the types of data that can be presented on maps. Weather maps (see the box below) and topological maps (see the box above) are among the common thematic maps.

M. NAGEL

See also: EARTH, STRUCTURE OF; REMOTE SENSING; SURVEYING; UNITED STATES GEOLOGICAL SURVEY.

Further reading:

McDonnell, P. *Introduction to Map Projections.* Rancho Cordova, California: Landmark, 1991. Monmonier, M. *Mapping It Out.* Chicago: University of Chicago Press, 1993. Tyner, J. *Introduction to Thematic Cartography.* Englewood Cliffs, New Jersey: Prentice-Hall, 1992.

WEATHER MAPS

Weather maps are valuable sources of information about the climate in a certain area. The first weather maps were made in the mid 1800s using the telegraphic collection of weather reports. They generally use standard symbols. Shading indicates cloud cover. Sometimes pictographs, such as snowflakes, are used to indicate the type of precipitation, if any, from the clouds. Barometric pressure is represented by connected lines of equal pressure called isobars, and the word *high* or *low* is printed at the center of these to indicate the type of pressure. Weather fronts are shown as darker lines, with triangles pointing in the direction that a cold front is moving and half circles pointing in the direction of a warm front. Some weather maps are produced to chart storms over a period of time, but most are obsolete within a day.

MARINE EXPLORATION

Marine exploration is the study of the underwater world

trapped at the top as it was dropped into the water. When the oxygen in the bubble was used up, the dive was over.

Compressed air systems were developed in the late 1700s, along with metal helmets and flexible suits. The cumbersome canvas suits and big copper helmets were a far cry from today's sleek wet suits, but they worked. Divers were tethered to a boat by a hose that supplied air and a lifeline that was used to raise them at the end of their dive.

In the 19th-century classic *Twenty Thousand Leagues Under the Sea*, Jules Verne (1828–1905) wrote of humans wearing portable breathing systems to explore the seas. His book was fiction, but, nearly a century later, French oceanographer Jacques-Yves Cousteau (1910–1997) developed a breathing system much like the one Verne had imagined. Today, thousands of amateur ocean explorers worldwide don masks, fins, and air tanks to explore the wonders of the deep.

Cousteau and partner Émile Gagnan revolutionized diving in 1943. Using an idea that Frenchman Yvies Le Prieu originated in 1933, Cousteau and Gagnan developed portable underwater breathing tanks, called aqualungs. Their tanks and similar apparatuses are commonly called scuba, for self-contained underwater breathing apparatus.

Recreational scuba divers can comfortably swim with the fish down to depths of about 165 ft (50 m). Professional divers have reached depths of 1970 ft (600 m) using highly pressurized air to counteract the water pressure. But this technique can be dangerous, because it can force toxic levels of inert gases present in the air into the bloodstream and may also cause decompression sickness (the bends).

The symptoms of the bends include acute pain in the muscles and joints, vomiting, deafness, faint-

CONNECTIONS

● Marine exploration often involves divers working at great depths in extremes of **TEMPERATURE** and pressure.

● **SATELLITES** are now being used to help explore and **MAP** the **OCEANS**.

Since ancient times, people have been fascinated by the mysteries of the deep. Records tell of divers in the Mediterranean Sea as far back as 4500 B.C.E. Many scientists believe that divers in 1300 B.C.E. first developed goggles from polished tortoiseshell to help them find pearls, sponges, and shells. A diver with powerful lungs could plunge up to 100 ft (30 m). But there his ability to search the sea ended.

According to legend, in 333 B.C.E. Alexander the Great (356–323 B.C.E.; king of Macedonia between 336 and 323 B.C.E.) was so curious about what lay beneath the surface of the sea that he commissioned a diving bell. It was a crude, barrel-shaped contraption, with a glass window for viewing fish and other sea life. The bell's only air supply was the bubble

CORE FACTS

■ Marine exploration is of growing importance because land resources are beginning to run out.

■ According to legend, the first diving bell was commissioned in 333 B.C.E. by Alexander the Great.

■ Sonar and, more recently, satellite surveys have made it possible to map the entire ocean floor.

■ Remote-controlled submersibles are being used to explore deep and narrow canyons.

■ Scuba diving enables humans to make close-up investigations of the marine environment.

■ The deepest part of any ocean, the Mariana Trench near the Philippines, was first reached by submersible in 1960.

ing, and paralysis. The disorder results when nitrogen in the blood forms tiny bubbles that block small blood vessels, preventing oxygenated blood from reaching the tissues. This potentially fatal condition can be avoided by replacing the pressurized air with a mixture of helium and oxygen.

Seamen begin scientific marine studies

From the middle of the 15th century onward, rival European sailors set out to find new lands and annex them for their own countries. A knowledge of the sea, and particularly of coastal waters in all parts of the globe, became of great strategic importance. Maps were so valuable that when Welsh pirate Henry Morgan (1635–1688) captured a book of charts from the Spaniards and presented it to the English king Charles II, he gained his pardon, a knighthood, and the governorship of Jamaica. During the 17th century, the Royal Society in England began to gather marine information: whenever a ship set sail on a long voyage, the captain was asked to keep a record of such physical characteristics as depth, salinity, and tidal flows wherever he went.

Italian geologist Luigi Ferdinando Marsili (1658–1730) made the first systematic study of the sea, and in 1724 he published his findings in his *Histoire Physique de la Mer*. He believed that land formations could not be understood without knowledge of formations under the sea. Through a series of soundings near Marseille, France, Marsili was able to draw an underwater map that correctly showed the continental shelf dropping off into deeper waters. He also analyzed the ocean's temperature, salinity, currents, and tides.

Navigation out of sight of land remained difficult. To navigate with any accuracy most sailors had to rely upon the position of the Sun at midday and the Moon and stars in the sky at night. However, an accurate clock is needed to keep time and the position of the Moon and the stars is often difficult to establish in a cloud-covered sky. Late in the 18th century, British naval captain James Cook (1728–1779) pioneered the use of the chronometer, which at last kept accurate time. This explorer took soundings all over the Pacific and made a great many significant contributions to marine exploration, including his finding in the Antarctic Ocean that warmer layers of water lie below the cold surface water.

The most famous early marine explorer in the United States was Benjamin Franklin (1706–1790). He conducted his studies during his eight voyages between America and Europe. His most important discovery was the Gulf Stream, the warm current that flows north along the east coast of the United States (see GULF STREAM). Captains immediately put Franklin's finding to use, considerably cutting their travel time between the United States and Europe.

While seamen gathered data about the oceans' physical features, marine plant and animal life remained obscure. One of the first people to study ocean life-forms was English naturalist Edward Forbes (1815–1854). In 1842, after failing to find evidence of life below 1800 ft (550 m), he proposed that the lower depths were an azoic ("without animals") zone. His theory held for more than 30 years before it was proven wrong.

English naturalist Charles Darwin (1809–1882) created a hunger for knowledge of the sea's creatures when he published *The Origin of Species* in 1859. In this revolutionary book, Darwin proposed that evolution occurred more slowly in the sea, where the environment was more stable than that of the land.

The first challenge to Forbes's azoic theory came in 1868 when Scottish scientist Charles Wyville Thomson (1830–1882) dredged sponges, crus-

Modern shallow-water diving often involves the use of self-contained underwater breathing apparatuses (scuba). This method of diving enables the individual to move freely in most underwater situations.

THE GERMAN ATLANTIC EXPEDITION

Although humans had been exploring the mysteries of the seas for thousands of years, it was not until the 1925 voyage of the vessel *Meteor* that marine exploration became a modern science. The two-year expedition was supported by the German government. Its mission was to investigate water circulation in the Atlantic Ocean.

Alfred Mertz and Georg Wüst led a scientific team of five physical oceanographers, two meteorologists, five chemists, a geologist, and a biologist. Their steamer was equipped with the latest current meters, echo sounders, plankton nets, and chemical laboratory equipment.

In two years, *Meteor* traveled more than 67,000 miles (some 110,000 km), collected more than 9000 temperature and salinity readings, and obtained more than 70,000 echo soundings. The soundings, for the first time, demonstrated the rugged face of the ocean floor. From the soundings, scientists mapped out the important Mid-Atlantic Ridge, a suboceanic rise running from Iceland to Antarctica. These underwater mountains had escaped the notice of the *Challenger* scientists, who 50 years earlier had conducted the most extensive marine research of that time.

HISTORY OF SCIENCE

The Deep-Star 4000, named **Westinghouse,** *is a research submersible that is used to take humans to depths of up to 4000 ft (1219 m). It is pictured here on the seabed in the Gulf of Mexico.*

taceans, and mollusks from depths greater than 1800 ft (550 m). Forbes's theory was finally disproved eight years later, following the historic voyage of the HMS *Challenger*. This British expedition was the most expensive and thorough scientific marine investigation made during the 19th century. Unlike earlier expeditions, its sole mission was to study the sea. During the ship's 3½-year circumnavigation of the globe, *Challenger* scientists made many important discoveries, including deep-sea life, a seafloor made of clay formed from the decomposition of skeletons of mollusks and other species, and manganese nodules on the floors of the Atlantic and Pacific Oceans (see MANGANESE). Specimens collected during the journey led to the identification of 700 new genera and 4000 new species.

Sonar maps out ocean's rugged landscape

The outbreak of World War I in 1914, with the rapid development of submarines, made marine research essential to military needs. The development of sonar by British and U.S. engineers in 1916–1918 revolutionized marine exploration.

Sonar consists of an underwater sound source and a microphone to pick up the sound waves as they return to the ship. A sound pulse is aimed toward the

Marine exploration often involves the discovery and study of shipwrecks. These divers are photographing a shipwreck off the north coast of France.

bottom, where it is reflected. In shallow water, the pulse returns quickly; in deep water, slowly. Hence, the water's depth is directly proportional to the time it takes the pulse to return.

Submarines using sonar equipment could continuously gauge the ocean's depth. Using sonar, scientists for the first time could create maps of the ocean floor. Those participating in the German Atlantic Expedition in 1925 (see the box on page 714) capitalized on the new technology to map out details of the Mid-Atlantic Ridge.

A new form of sonar captures "snapshots" of submerged objects without emitting a sound. Instead, an acoustic daylight ocean noise imaging system (ADONIS), developed in 1992, uses submerged microphones to pick up the ocean's random noise. This noise bounces off objects, just as ambient light does above the surface. Like a camera picking up reflected light, this imaging system picks up reflected sound. Underwater electronics relay the information to a computer on the ship, where it is translated into an image.

To map all Earth's oceans by sonar would take a whole fleet many years. Instead, researchers are turning to satellites to fill in the gaps. Satellites orbiting Earth bounce microwaves off the surface of the ocean, much like sonar bounces sound waves off the ocean floor. Because the surface follows the laws of gravitational fields, it dips over the ocean's canyons and swells over its mountains. Although a sailor could

AUGUST AND JACQUES PICCARD

In 1960, seven years after Sir Edmund Hillary (1919–) scaled Mount Everest, U.S. Navy lieutenant Don Walsh and Swiss engineer Jacques Piccard (1922–) plunged to the bottom of the "Everest of the Seas." Their pioneering voyage took them to the ocean's greatest depth, the 7-mile- (11-km-) deep Challenger Deep—the deepest point of the Mariana Trench, a subduction zone (see PLATE TECTONICS) near the Philippine Islands. As they peered through the windows of their bathyscaphe, the men came eye to eye with a flounder-like fish, swimming contentedly in water under a pressure of 17,000 lb per sq in (1150 atmospheres). For Piccard, this risky endeavor fulfilled his destiny. He was born into a family of scientists drawn to the limits of sea and sky.

His father, physicist Auguste Piccard (1884–1962), first came up with the idea for the bathyscaphe as a student at Zurich Polytechnic School around 1920. Piccard learned there of problems researchers had encountered in studying fish netted from the ocean's greatest depths. Every time the fish were hauled to the surface, the higher temperature and lower pressure killed them and extinguished their unusual phosphorescing organs. Piccard realized that the only way to examine these animals alive would be in their habitat. He decided there must be a way to build a chamber with portholes that could resist the underwater pressure.

The young physicist, however, set his idea aside while he pondered the limits of flight in hot-air balloons. Balloon pilots had been limited by the upper atmosphere's cold, low pressure, and lack of oxygen. Piccard overcame those limits by building an air-tight aluminum gondola that could be pressurized. In 1931, Piccard floated up to almost 10 miles (16 km), higher than humans had ever climbed.

Having conquered the skies, Piccard then turned his attention back toward the seas. He used his knowledge of pressurized chambers in 1937 to design the first bathyscaphe. In 1953, he descended to a record depth of 10,000 ft (3000 m). It was his bathyscaphe *Trieste* that, seven years later, took son Jacques on his monumental plunge into the Mariana Trench.

DISCOVERERS

never see it, the surface of the ocean deviates up to 100 ft (30 m). In 1995, satellite images revealed an unusually large swelling 600 miles (960 km) northwest of Easter Island. When researchers sailed to the site for a closer look, they discovered the greatest concentration of active volcanoes on Earth.

Oceanography institutions take over research

Today, information about the oceans is gathered on board large research ships operated by oceanographic institutes. Scuba pioneer Cousteau, with his vessel *Calypso*, remains among the world's foremost researchers. Prominent research centers include the Woods Hole Oceanographic Institution in Massachusetts and the Scripps Institution of Oceanography in California.

Research vessels have come a long way since HMS *Challenger*. Satellite navigation, combined with computers, can pinpoint a ship's location with far greater accuracy than yesteryear's chronometers.

Key components of the modern research ships are their laboratories. Wet labs on deck provide a work area to sort and clean large specimens. Samples may be analyzed later in a chemistry laboratory below. Modern ships also have an electronics laboratory, filled with computerized apparatus to measure and analyze such physical properties as magnetic fields, the thickness of Earth's crust, and underwater sounds.

Submersibles carry researchers into the deep

Curiosity about sea life in 1929 inspired U.S. naturalist Charles William Beebe (1877–1962) to build a round diving chamber dubbed the bathysphere (from the Greek word *bathos*, meaning "depth"). His invention held two men and looked like a large ball with two portholes. He chose a spherical shape because spheres resist pressure evenly.

Beebe and partner Otis Barton took their maiden voyage on June 3, 1930, near Bermuda. Their bathysphere was lowered by cables from a ship to a depth of 200 ft (60 m). Over the next four years, Beebe made several dives in the bathysphere. His most spectacular was on August 15, 1934, when he reached a record-breaking depth of 3028 ft (908 m).

In 1953, Swiss physicist August Piccard (1884–1962) shattered this record by plunging to a depth of 10,000 ft (3000 m) in a bathyscaphe he had designed (see the box on page 715). Then, in 1960, his son Jacques Piccard (1922–) and U.S. Navy lieutenant Don Walsh made history when they took the elder Piccard's bathyscaphe *Trieste* down 7 miles (11 km), at the ocean's deepest point, in the Mariana Trench near the Philippine Islands. Their dive revolutionized marine exploration. Suddenly, every research institute wanted a submersible.

In 1959, Cousteau developed a saucer-shaped vehicle called the *Soucoupe Plongeante* (French for "diving saucer"). Then, in 1964, Allyn Vine devised a mid-depth submersible for Woods Hole Oceanographic Institution. His vehicle, named *Alvin*, was modeled after *Trieste*. But, unlike its predecessor, *Alvin* was fitted with scientific instruments such as salinity, temperature, and depth recorders. Cameras and vacuum attachments to collect fish made it particularly valuable to marine researchers.

The 1970s ushered in a movement toward self-contained atmospheric diving suits (or SCADS). These articulated metal suits enabled divers to walk on the seabed up to depths of about 2000 ft (600 m).

The internal layout of the first submarine propelled by an electric generator. It was tested in Cherbourg, France, in 1889 and remained submerged for about eight hours.

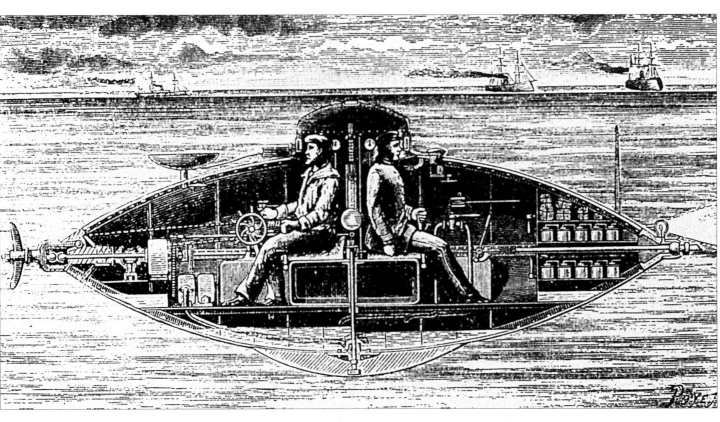

Small untethered vehicles developed in the late 1970s and 1980s are able to explore lakes and other areas too small for a large tending ship. A popular one called *Deep Rover* is a clear acrylic sphere with arms the pilot can maneuver like hands. *Deep Rover* is said to be as easy to operate as a golf cart.

Submersibles traditionally descend using weights that are later left behind or water that is displaced with air. A new generation of one-person, battery-powered vehicles disposes with this ballast system to "fly" through the water. The high-speed *Deep Flight* was designed for such innovative research as swimming alongside dolphins and whales, among a school of fish, or through underwater canyons too narrow for traditional submersibles.

The deep canyons that lured early bathyscaphe designers increasingly are left to unpiloted submersibles. Their small size gives these aquatic robots an advantage in narrow canyons. They also operate quicker and at less cost than most piloted vehicles.

In 1993, the Japanese unpiloted bathyscaphe *Kaiko* ventured into the Challenger Deep—the deepest point of the Mariana Trench, which is located near the Philippine Islands. This submersible was operated by fiber-optic cable from a nearby ship, providing control of five television cameras and two arms with which samples weighing up to 65 lb (30 kg) could be collected. Unpiloted submersibles of this type have applications in a number industries and are operated by other research institutes, oil-prospecting companies, and television production companies that make underwater movies.

The sea may be the solution to problems on land

No longer the exclusive laboratory of oceanographers, the ocean is being closely examined by pharmacologists, nutritionists, gas and oil prospectors, and other researchers seeking solutions to problems on land.

Dr. Robert Ballard (1942–), who made headlines when he found the RMS *Titanic* in 1985, believes that land is running out for Earth's ever increasing population. With more than 70 percent of Earth's surface covered with water, he envisions roomy new human houses aboard artificial islands on the ocean's surface. The first step in his vision would involve gathering more information about the oceans to ensure that his proposal is viable. Ballard proposes a team of submerged robotic vehicles combing the ocean floors and gathering information for humans' buoyant future.

Other researchers are looking to the seas for relief of food shortages. Projects under study include increasing fish catches through sonar and chemical bait. The ocean one day may also provide potable water, as methods of removing salt improve. Pharmacologists are searching for new drugs under the sea. Hundreds of chemicals that ward off marine predators are being tested for their ability to defend against human viruses, fungi, and cancer.

Another potential gift from the sea is energy. Researchers are seeking a way to tap into the power of the waves, tides, and the heat energy from the ocean's temperature differences. If successful nuclear fusion develops (see FUSION), the hydrogen-rich water present in every ocean could provide an unlimited supply of deuterium (see HYDROGEN).

Military submersibles, such as the USS Lafayette are the largest submersibles and use the most advanced technology. They need detailed marine data when maneuvering beneath the surface.

DISCOVERING SHIPWRECKS

Like California's gold mines, sunken ships that were once used to transport valuables around the world have lured people with dreams of striking it rich. But until the relatively recent introduction of scuba equipment and sonar, their chances were about as good as winning the lottery. Nevertheless, as early as 1685, American Captain William Phips (1651–1695)—who later was appointed first governor of Massachusetts—succeeded in recovering the treasure of the Spanish ship *Almiranta*, which had sunk off the coast of Cuba in 1641. His divers used a Bermuda tub, a barrel full of air that was weighted so that it would sink; they swam to it to take in air.

As scuba diving became popular in the decades following World War II, amateur divers began turning up rusty canons and other artifacts on the French and Italian Rivieras and off the Florida Keys. At about the same time, marine archaeology began to develop. Students of the budding science sought to preserve and reconstruct the history of sunken ships, rather than seek for hoards of treasure.

The first U.S. venture into underwater archaeology came in 1960, when a team from the University of Pennsylvania Museum excavated a wreck from 1200 B.C.E. near Cape Gelidonya, Turkey. The skilled crew members included draftsmen, photographers, and archaeologists. They pieced together the history of the cargo ship's last voyage.

Today, the search for shipwrecks depends on fiber-optic cables, sophisticated underwater cameras, and robots. Such technology led to the monumental discovery of the RMS *Titanic* in 1985, 73 years after it sank in the icy North Atlantic.

Dr. Robert Ballard of Woods Hole Oceanographic Institution and Jean-Louis Michel of the French National Institute of Oceanography led the expedition. But it was a sledlike assembly of sonar equipment and moving cameras named *Argo* that actually found the luxury liner's wreckage. The men watched video screens as *Argo*, towed from their ship *Knorr*, brought it into view.

A year later, Ballard returned with a crew from Woods Hole to explore the ship. He and two colleagues descended 2½ miles (4 km) to the ship's deck in the submersible *Alvin*. From there they sent a small, tethered robot called *Jason Junior* on a photography mission down the grand staircase into the belly of the ship.

Marine archaeologists and salvagers increasingly are relying on remotely operated vehicles like *Jason Junior*. U.S. salvager Tom Thompson and his Columbus America Discovery Group in 1986 found the 130-year-old, gold-laden wreck of the SS *Central America* using a sonar technology. The two submerged sonar transducers operate at low frequencies to cover a wide range.

Sophisticated techniques have enabled scientists to discover many shipwrecks such as the **Santa Teresa di Gallura,** *off the coast of Sardinia, Italy.*

A CLOSER LOOK

An abundance of methane gas trapped over millions of years in water crystals formed by intense pressure was recently discovered off the coast of North Carolina. Scientists estimate that combustion of this source of methane has the potential to provide all of the world's energy needs for 20 years. But recovering the gas will not be easy. One suggested approach is to circulate warm surface water through a pipe to melt the buried crystals and liberate the gas. However, some researchers are concerned that the technique might weaken pockets in the ocean floor and cause them to collapse on the drilling equipment. The delicate challenge of drilling for the gas began late in 1995 with a team representing 16 nations.

Scientists dig deeper below the oceans

Using advanced drilling techniques and sonar, scientists are probing deep beneath the sea. Seafloor samples extracted by drills have revealed productive oil pockets and provided a history of Earth's climate, and the detailed work by oceanographers provided evidence to develop the theory of plate tectonics (see PLATE TECTONICS).

In 1995, researchers from the Woods Hole Oceanographic Institution used a 2-mile- (3.2-km-) long drill to probe the underwater equivalent of Old Faithful (a geyser in Yellowstone National Park; see GEYSERS). The researchers anchored about 2000 miles (3200 km) east of Miami, where a mound the size of the Houston Astrodome periodically belches hot, sulfurous plumes into the water. The mound was formed from the geyser's minerals, which precipitate when they reach the cool waters of the deep ocean. Marine explorers in recent decades have found similar mounds scattered in the Atlantic and Pacific Oceans.

Another tool for taking a look at the seafloor is sonar. Because sound travels at different speeds through different materials, scientists can determine whether the material below the ocean is hard rock, for example, or a magma pool. In 1995, this technology enabled researchers from the Scripps Institution of Oceanography to discover a magma pool along the submerged East Pacific Rise volcano range. The magma is expected to erupt in one of the planet's most spectacular underwater volcanoes. To be sure they do not sleep through the action, researchers installed hydrophones in the area to pick up the volcanoes' first rumblings.

C. WASHAM

See also: CONTINENTAL SHELVES; GEYSERS; OCEANOGRAPHY; OCEANS AND OCEAN ABYSSES.

Further reading:
Ballard, R. D. and McConnell, M. *Explorations: My Quest for Adventure and Discovery Under the Sea.* New York: Hyperion, 1995.
The Encyclopedia of the Earth: Oceans and Islands. Edited by F. H. Talbot and R. E. Stevenson. McMahons Point, New York: Smithmark Publishers, 1991.

INDEX

NOTE: **Italic numbers indicate illustrations and/or captions.**